Water-in-Plants Bibliography

volume 5 1979

References no. 5249–6605 / ACO–ZYA

Editors J. Pospíšilová and J. Solárová

Springer-Science+Business Media, B.V. 1981

Contributors

J. Solárová
J. Pospíšilová
Z. Šesták
J. Čatský
I. Tichá
D. Hodáňová

ISBN 978-90-6193-905-4   ISBN 978-94-017-5469-9 (eBook)
DOI 10.1007/978-94-017-5469-9

PREFACE

The fifth volume of Water-in-Plants Bibliography includes papers in all fields of plant water relations research which appeared during the year 1979 - from theoretical considerations about the state of water in cells and its membrane transport to drought resistance of plants or physiological significance of irrigation. In addition to papers devoted entirely to plant water relations, papers on other topics are included if they contain data on plant hydration level, water vapour efflux, rate of water uptake or water transport, etc., or if they contain valuable methodological information (measurement of selected microclimatic factors, soil moisture, etc.).

We have tried to cover fully the relevant papers which have been published in the most important scientific periodicals and books. Articles appeared in local journals, mimeographed booklets, abstracts of thesis and of symposia contributions, etc., were chosen mostly from reprints received directly from authors. The courtesy of those authors who have already supplied us with reprints and lists of their publications is highly appreciated. The manuscript is usually prepared in May and June of the year following the year which it covers. Unfortunately some reprints come later and thus the respective references appear in the following volume, with one year delay.

To maximize the value of the bibliography the references are arranged alphabetically according to the authors' names, and each volume is provided with three indexes. The authors' index contains all names of authors, co-authors and editors. Plant genera used as experimental material are indexed according to their Latin names. The subject index covers primary items chosen according to the interest of water relations researchers. Its preparation was based not only on the titles, key words and abstracts but also on the whole content of the article.

This volume is accompanied with cumulative indexes to volumes 1 - 5.

Since more than 1500 relevant papers dealing with plant water relations and relative topics are published every year and included in this bibliography, and since all citations have been checked with the originals, collecting and preparing for publication such a large amount of material would have been impossible without the collaboration of our colleagues from the Department of Physiology of Photosynthesis and Water Relations of the Institute of Experimental Botany of the Czechoslovak Academy of Sciences in Prague. We have also acknowledge with thanks the cooperation of Mrs. Ludmila Hávová, Mrs. Lenka Kolčabová, Mrs. Marie Mandlová, Mrs. Marta Smídová and Mrs. Drahomíra Těžká who helped in typing card material and Mrs. Zora Zawoyská and Mr. Petr Zázvorka who supplied us with rare periodicals.

Dr. Jana Pospíšilová and Dr. Jarmila Solárová

Institute of Experimental Botany
Czechoslovak Academy of Sciences

Flemingovo nám. 2
160 00 PRAHA 6
Czechoslovakia

Praha, 30 October 1980

INSTRUCTIONS FOR USE

All references are arranged alphabetically according to the authors' names. They are numbered and these numbers are used in the indexes. An asterisk preceding the number denotes the reference published in the preceding period (1975 - 1978).

Authors' names are presented in the spelling used in the original paper. If this spelling does not correspond to the spelling usually used by the author (e.g. Russian papers of English authors), one spelling is referred to the other in the Authors' Index. Like the transcriptions they are alphabetically arranged mostly according to the authors' own references. Nevertheless, the editors apologize for some misinterpretations which are partly corrected by the cross-indexing in the Authors' Index.

The references contain the original unshortened title of the paper (book). English, French, and German titles are cited in the original language. Titles in other languages are supplemented with a translation in English (using the title of the respective English abstract, if it is presented). Titles of Japanese, Chinese etc. papers are given in English translation only. In both these cases the abbreviations of the original language and the language of the abstract are given in brackets at the end of the reference. The following abbreviations are used most frequently:

Belorussian	Japanese
Bulgarian	Latvian
Chinese	Lithuanian
Croatian	Norwegian
Danish	Polish
English	Russian
Esthonian	Roumanian
French	Slovak
German	Spanish
Georgian	Swedish
Hungarian	Ukrainian
Italian	Uzbeg

The transliteration of Cyrillic characters is in accordance with the BSI-ASA/SC-Z39 draft table, i.e.:

a	а		p	п
b	б		r	р
ch	у		s	с
d	д		sh	ш
e	е		shch	щ
ė	э		t	т
f	ф		ts	ц
g	г		u	у
i	и		v	в
ĭ	й		y	ы
k	к		ya	я
kh	х		yu	ю
l	л		z	з
m	м		zh	ж
n	н		"	ъ
o	о		'	ь

Several exceptions apply for Ukrainian and Belorussian:

Ukrainian:	y	и
	i	і
	ĭ	ї
Belorussian:	ŭ	ў

The journals' names are abbreviated mainly according to the Style Manual for Biological Journals (Second Edition, Amer. Institute of Biological Sciences, Washington, D.C. 1964), e.g.:

Abhandlungen
Abstract
Abteilung
Academy
Acker
Acta
Advances
Africa (-ican)
agricultural
Agriculture
Agrobiology (-ogiya)
Agrobotanica
Agrokémia
Agronomy
agropecuaria
Akademie (-emiya)
Algology
allgemeine
Amélioration
America
American
Anais (-alele)
Analysis
analytical
Anatomy
angewandte
animal
Annales (-als)
annual
anorganic (-anisch)
applied
aquatic
Arbeit
Archiv
Argentina
Association
Atmosphere
atmospheric
atomic
Australia (-ralian)
Azerbaĭdzhanskaya
Bacteriology
Beiheft
Beiträge
Belgique
Belorusskaya
Berichte
biochemical
Biochemie
Biochemistry
biochimica
biokhimicheskiĭ
Biokhimiya
Bioklimatologie
Biologia (-ogy)
biological (-ogisk)
biophysical
Biophysics
Bodenkunde
Boletin (-ettino)
Bolgarskiĭ
botanica (-anicorum)
botanical (-anisca)
Botanika (-any)

Brasileira
Brazil
Breeding
British
Bulletin (-etins)
Byulleten
California
Canada (-adian)
cellular (-ulaire)
Center
central
Centralblatt
Československý
chemical
Chemistry
chimicus
Chinese
Chromatography
Chronicle
Ciencia
cientificas
College
Commision
Communication
comparative
Comptes Rendus
Conference
Congress
Conservation
Contamination
Contribution
Control
Croatica
cultural
Culture
current
Cytobiology
Cytochemistry
Cytology
Czechoslovak
Danske
dendrological
Dendrology
Department
Deutsche (-schland)
Development
Disease
Dissertation
Division
Doklady
Dopovidi
Drainage
ecological
Ecology
Economy
Edafology
Education
Ékologiya
eksperimental'nyĭ
Embryology
Encyclopedia
Engineering
Enology
Entomology

environmental
Enzymology
Estonskaya
European
Experiment
experimental
Faculty
Federation
Fizika
Fiziologiya
Flurbereinigung
forestiere
Forestry
Forschung
Foundation
France
Gazette
general
genetical
geneticheskiĭ
Genetics (-ika)
Geobotany
Geofizika
Geophysics
Gesellschaft
Giornale
gosudarstvennyĭ
Government
Grassland
Gruzinskaya
Helveticus
Histochemistry
Histoire (-ory)
Histology
horticultural
Horticulture
Hungaricae
Hungaricus
Husbandry
Hydrobiology
Hydrology
Indian
Industry
inorganic
Institute
Institutului
international
Investigation
Irrigation
Isotopes
issledovatel'skiĭ
Italian (-y)
Izvestiya
Jahrbuch
Japan (-anese)
Journal
Khimiya
Klasse
Kongelige
Közlemenyek
kul'turnykh
Laboratory
Landbauforschung
Landwirtschaft

lesní (-ího)
Letters
Limnology
Linnean
Litovskoĭ
Lucrarile
Magazin
Management
marina (-ine)
Material
Mathematics
Mededelingen
mediterranean
Meldinger
Meteorology
Microbiology
Midland
Mitteilungen
Modeling
modern
molecular
Monographiae (-aphy)
Moskovskiĭ (-ovskogo)
Mycology
national
natural
Naturalist
naturelle
naturkundliche
Naturforschung
nauchnye (-nyĭ)
Neerlandica
Netherland
New Zealand
Norges
Norwegian
Notiser
nuclear
Nutrition
obshcheĭ (-iĭ)
Oceanography
Oecologia
Ökologie
Optics
opytnaya (-yĭ)
organic
original
ornamental
Otdelenie
Paleobotany
Palynology
Pathology
pedagogicheskiĭ
Pesquisa
Pesticide
Pflanzen-
Pflanzenernährung
Pflanzenphysiologie
Pflanzenzüchtung
Philosophy
Photogrammetric
Phycology
physical
Physics

physiological	rolniczych	SSSR	Ukrains'kaya
Physiology	Rostlin (-lina)	Stantsii (-ntsiya)	Universidad (-ersity)
Phytologist	rostlinná	Station	US, USA
Phytopathology	Roumaine	stiintifice	USSR
Phytotaxonomy	royal	subtropicale	Uzbekskiĭ (-ekskaya)
Plantarum	Russian	summary	vědecké (-ecký)
Polonica (-ska)	Russkiĭ	Supplement	vegetable
Pollution	Sbornik	Survey	végétale
Práce	Scandinavica	Swedish	Verhandlungen
Practice	Scandinavicus	Symposium	Veröffentlichungen
prikladnoĭ	School	System	Vestnik
Proceedings	Science	Tagungsberichte	Videnskabernes
Progress	scientific	technical (-nische)	Virology
Publication	Section	Technology	Virusforschungen
Publishers	Selektsiya	Tekhnika	Viticulture
Quality	Selskabs	theoretical	Volume
quantitative	Sel'skokhozyaĭstvo	thermal	Voprosy
Quarterly	Series (-iya)	Tidsskrift	vostochnyĭ
Radiation	Service	Tijdschrift	vsesoyuznyĭ
Radiobiology	Shkoly (-oly)	Toxicology	vyssheĭ (-iĭ)
Rasteniĭ	Sibirskiĭ (-skogo)	Transactions	výzkumný (-umného)
Rastenievodstvo	Skrifter	Travail (-aux)	Weekblatt
Recherche (-erches)	Slovak (-enská)	tropical (-icale)	Wetenschappen
Report	Society	Trudy	Wissenschaft
Research	Soobshcheniya	Turkmenskaya	Zapiski
Resources	Sovetskiĭ (-iet)	uchenye	Zeitschrift
Review (-ista, -ue)	sovremennyĭ	Ugeskrift	Zeitung
Rivista	special	United Kingdom	Zentralblatt
Roczniky	sperimentale	Ukraĭnian	Zhurnal

The numbers at the end of each reference of a journal article denote: volume (issue) : first page - last page, year of publication. The number of issue is given only in journals where each issue is paginated separately.

Book titles are cited according to the title page, not to the book jacket or cover. The publishing house, place and year of publication are included.

Printers' errors in the original papers are marked by underlining the respective words (letters).

5249 - ACOCK, B., NICHOLS, R.: Effects of sucrose on water relations of cut, senescing, carnation flowers. - Ann. Bot. 44: 221-230, 1979.

5250 - ADEPIPE, N.O., TINGEY, D.T.: Ozone phytotoxicity in relation to stress ethylene evolution and stomatal resistance in cowpea (*Vigna unguiculata*) cultivars. - Z. Pflanzenphysiol. 93: 259-264, 1979.

*5251 - ADJAHOSSOU, F., VIEIRA da SILVA, J.: Teneur en glucides solubles et en amidon et résistance à la sécheresse chez le palmier à huile. - Oléagineux 33: 599-603, 1978.

5252 - AGARWAL, S.K., DE, R.: Effect of nitrogen rates, mulching and antitranspirants on water use and water use efficiency of barley (*Hordeum vulgare* L.) varieties grown under dryland conditions. - J. agr. Sci. 92: 263-268, 1979.

*5253 - AGRAWAL, P.K.: Changes in germination, moisture and carbohydrate of hexaploid triticale and wheat (*Triticum aestivum*) seeds stored under ambient conditions. - Seed Sci. Technol. 6: 711-716, 1978.

5254 - AHLGRIMM, H.-J.: Der Einfluss verschiedener Witterungsfaktoren auf die Entwicklung von Grashalmen (*Phleum pratense*) und ihre Wirkung auf die Halmfestigkeit. - Angew. Bot. 53: 161-173, 1979.

5255 - AHMAD, I., LARHER, F., STEWART, G.R.: Sorbitol, a compatible osmotic solute in *Plantago maritima*. - New Phytol. 82: 671-678, 1979.

5256 - AHMAD, K.J.: Stomatal features of *Acanthaceae*. - In: SEN, D.N., CHAWAN, D.D., BANSAL, R.P. (ed.): Structure, Function and Ecology of Stomata. Pp. 43-60. Bishen Singh Mahendra Pal Singh, Dehra Dun 1979.

*5257 - AHMAD, K.J.: Epidermal studies in *Fittonia* Coemans (*Acanthaceae*). - Feddes Repertorium 89: 369-374, 1978.

5258 - AHMAD, K.J.: Taxonomic significance of epidermal characters in *Acanthaceae*. - Progress Plant Res. 1: 135-160, 1979.

5259 - AHMED, A.M., HEIKAL, M.M., SHADDAD, M.A.: Changes in some plant-water relation parameters of some oil producing plants over a range of salinity stresses. - Biol. Plant. 21: 259-265, 1979.

5260 - AHMED, A.M., HEIKAL, M.M., SHADDAD, M.A.: Growth, photosynthesis and fat content of some oil producing plants as influenced by some salinization treatments. - Phyton 19: 259-267, 1979.

5261 - AHO, N., DAUDET, F.-A., VARTANIAN, N.: Évolution de la photosynthèse nette et de l'efficience de la transpiration au cours d'un cycle de dessèchement du sol. - C.R. Acad. Sci. Paris, Sér. D 288: 501-504, 1979.

5262 - AKITA, S., TANAKA, I.: Studies on the mechanism of differences in photosynthesis among species V. Stomatal response in high oxygen concentration and its effect on the rate of apparent photosynthesis. - Jap. J. Crop Sci. 48: 470-474, 1979.

5263 - ALBERGONI, F.G., BASSO, B., TOSO, S.: Considerations on CO_2 effluxes in *Zea mays* L. and *Trifolium repens* L. - Maydica 24: 113-124, 1979.

5264 - ALEKSIDZE, G.N., RACHVELISHVILI, É.V., POTSKHVERIYA, A.M.: K izucheniyu vredonosnosti plodovykh listovykh tlei v Gruzii. [Damage caused by fruit aphids in Georgia.] - Soobshch. Akad. Nauk. Gruz. SSR 93(1): 193-195, 1979. [In R, ab: E, Georg.]

*5265 - ALLEN, L.H.,Jr., LEMON, E.R.: Carbon dioxide exchange and turbulence in a Costa Rican tropical rain forest. - In: MONTEITH, J.L. (ed.): Vegetation and the Atmosphere. Vol. 2. Case Studies. Pp. 265-308. Academic Press, London - New York - San Francisco 1976.

5266 - ALSTON, A.M.: Effects of soil water content and foliar fertilization with nitrogen and phosphorus in late season on the yield and composition of wheat. -
Aust. J. agr. Res. 30: 577-585, 1979.

5267 - AMBERGER, A.: Pflanzenernährung. - Ökologische und Physiologische Grundlagen
(Uni-Taschenbücher 846). - Verlag Eugen Ulmer, Stuttgart 1979.

*5268 - ANANYAN, A.A., EGIAZARYAN, A.G., SARKISYAN, E.M.: Fiziologo-biokhimicheskaya
kharakteristika sortov tomatov razlichnoĭ skorospelosti. [Physiological and
biochemical charakteristics of tomato cultivars ripening at different times.]
- Biol. Zh. Armenii 28(6): 66-69, 1975. [In R.]

5269 - ANDERSEN, A.S.: Plant growth retardants: Present and future use in food production. - In: SCOTT, T.K. (ed.): Plant Regulation and World Agriculture. Pp.
251-277. Plenum Press, New York - London 1979.

5270 - ANDERSON, G.W., BATINI, F.E.: Clover and crop production under 13- to 15-year-
old *Pinus radiata*. - Aust. J. exp. Agr. anim. Husb. 19: 362-368, 1979.

*5271 - ANDERSON, W.K., SMITH, R.C.G., McWILLIAM, J.R.: A systems approach to the
adaptation of sunflower to new environments. II. Effects of temperature and
radiation on growth and yield. - Field Crop Res. 1: 153-163, 1978.

5272 - ANDONOVA, P., MEKHANDZHIEVA, A.: Nyakoi promeni v plastidnite pigmenti na lyutsernata pod vliyanie na razlichnata vodoobezpochenost i nachini na napoyavane.
[Some changes in alfalfa plastid pigments resulting from varying water supply
and irrigation method.] - Fiziol. Rast. (Sofia) 5(1): 33-42, 1979. [In Bulg,
ab: R,E.]

5273 - ANDRÉ, M., DAGUENET, A., MASSIMINO, J., MASSIMINO, D., RICHAUD, C.: Le labora-
toire $C_2$3A. Un outil au service de la physiologie de la plante entière II. -
Possibilités de la mini-informatique et premiers résultats. - Ann. agron. 30:
153-166, 1979.

5274 - ANDRÉ, M., GERBAUD, A.: Consommation d'oxygène pendant la photosynthèse chez
Zea mays. - C.R. Acad. Sci. Paris, Sér. D 289: 793-796, 1979.

*5275 - ANGELOV, É.: Sezonnaya dinamika sostava lizimetricheskikh vod v seroĭ lesnoĭ
pochve Bolgarii. [Seasonal dynamics of the lysimetric water composition in a
grey forest soil of Bulgaria.] - Pochvovedenie 1978(3): 35-41, 1978. [In R, ab:
E.]

*5276 - AOTA, S., HOSHINO, M.: [Yearly variations of corn yield by conversion from
paddy field of ill-drained clayey soil into upland field.] - J. Jap. Soc. Grassland Sci. 24: 118-122, 1978. [In Jap, ab: E.]

5277 - ARDITTI, J.: Aspects of the physiology of orchids. - Adv. bot. Res. 7: 421-655,
1979.

5278 - ARMSTRONG, W.: Aeration in higher plants. - Adv. bot. Res. 7: 225-332, 1979.

5279 - ASHENDEN, T.W.: Effects of SO_2 and NO_2 pollution on transpiration in *Phaseolus
vulgaris* L. - Environ. Pollut. 18: 45-50, 1979.

5280 - ASHTON, D.H., TURNER, J.S.: Studies on the light compensation point of *Eucalyptus regnans* F.Muell. - Aust. J. Bot. 27: 589-607, 1979.

5281 - ASHWORTH, L.J., Jr., HUISMAN, O.C., HARPER, D.M., STROMBERG, L.K.: Verticillium
wilt disease of tomato: Influence of inoculum density and root extension upon
disease severity. - Phytopathology 69: 490-492, 1979.

5282 - ASHWORTH, L.J., Jr., HUISMAN, O.C., HARPER, D.M., STROMBERG, L.K., BASSETT, D.M.:
Verticillium wilt disease of cotton: Influence of inoculum density in the
field. - Phytopathology 69: 483-489, 1979.

5283 - **ASTON, M.J., LAWLOR, D.W.:** The relationship between transpiration, root water uptake, and leaf water potential. - J. exp. Bot. 30: 169-181, 1979.

5284 - **ATKINSON, D., WILSON, S.A.:** The root-soil interface and its significance for fruit tree roots of different ages. - In: HARLEY, J.L., SCOTT RUSSELL, R. (ed.): The Soil-Root Interface. Pp. 259-271. Academic Press, London - New York - San Francisco 1979.

*5285 - **AUCLAIR, A.N.D., BOUCHARD, A., PAJACZKOWSKI, J.:** Plant standing crop and productivity relations in a *Scirpus-Equisetum* wetland. - Ecology 57: 941-952, 1976.

5286 - **AUGUSTINE, J.J., STEVENS, M.A., BREIDENBACH, R.W.:** Physiological, morphological, and anatomical studies of tomato genotypes varying in carboxylation efficiency. - J. Amer. Soc. hort. Sci. 104: 338-341, 1979.

5287 - **AUSSENAC, G., GRANIER, A.:** Etude bioclimatique d'une futaie feuillue (*Fagus silvatica* L. et *Quercus sessiliflora* Salisb.) de l'Est de la France. II. - Etude de l'humidité du sol et de l'évapotranspiration réelle. - Ann. Sci. forest. 36: 265-280, 1979.

5288 - **AVERY, D.J., PRIESTLEY, C.A., TREHARNE, K.J.:** Integration of assimilation and carbohydrate utilization in apple. - In: MARCELLE, R., CLIJSTERS, H., Van POUCKE, M. (ed.): Photosynthesis and Plant Development. Pp. 221-231. Dr. W. Junk bv Publishers, The Hague - Boston - London 1979.

5289 - **AYRES, P.G.:** CO_2 exchanges in plants infected by obligately biotropic pathogens. - In: MARCELLE, R., CLIJSTERS, H., Van POUCKE, M. (ed.): Photosynthesis and Plant Development. Pp. 343-354. Dr. W. Junk bv Publishers, The Hague - Boston - London 1979.

5290 - **AYRES, P.G., ZADOKS, J.C.:** Combined effects of powdery mildew disease and soil water level on the water relations and growth of barley. - Physiol. Plant Pathol. 14: 347-361, 1979.

5291 - **BACHELOR, J.T., SCOTT, H.D.:** Effects of irrigation on nitrogen and potassium uptake by soybeans. - Arkansas Farm Res. 28(2): 5, 1979.

5292 - **BAIER, W.:** Note on the terminology of crop-weather models. - Agr. Meteorol. 20: 137-145, 1979.

5293 - **BAKER, E.A., BUKOVAC, M.J., FLORE, J.A.:** Ontogenetic variations in the composition of peach leaf wax. - Phytochemistry 18: 781-784, 1979.

5294 - **BAKER, R.L., POWELL, J.:** Western Oklahoma sandhill prairie yield and crude protein response to atrazine, nitrogen, and 2,4-D during drought. - In: GOODIN, J.R., NORTHINGTON, D.K. (ed.): Arid Lands Conference on Plant Resources. Pp. 564-573. Texas Technical University, Lubbock 1979.

5295 - **BALDY, C.M.:** Utilisation d'une relation simple entre le bac classe A et la formule de Penman pour l'estimation de l'ETP en zone soudano-sahélienne. - Ann. agron. 29: 439-452, 1978.

*5296 - **BALL, D.M., HOVELAND, C.S.:** Alkaloid levels in *Phalaris aquatica* L. as affected by environment. - Agron. J. 70: 977-981, 1978.

5297 - **BALLARD, J.T.:** Fluxes of water and energy through the Pine Barrens ecosystems. - In: Climate, Water, and Aquatic Ecosystems. Pp. 133-146. Academic Press, London - New York - San Francisco 1979.

5298 - **BALNOKIN, Yu.V., STROGONOV, B.P., KUKAEVA, E.A., MEDVEDEV, A.V.:** Zashchitnaya funktsiya membran kletok *Dunaliella* pri vysokikh kontsentratsiyakh NaCl v srede. [Protective function of *Dunaliella* cell membranes under high NaCl concentrations in the medium.] - Fiziol. Rast. 26: 552-559, 1979. [In R, ab: E.]

*5299 - BAŇOCH, Z., MATĚJÍKOVÁ, O.: Příspěvek k zakládání porostu jetelovin v závlaho-
        vých podmínkách. [Establishment of a clover stand in an irrigated area.] -
        Rost. Výroba (Praha) 24: 1135-1144, 1978. [In Czech, ab: R,E,G.]

 5300 - BARBER, S.A.: Growth requirements for nutrients in relation to demand at the
        root surface. - In: HARLEY, J.L., SCOTT RUSSELL, R. (ed.): The Soil-Root Inter-
        face. Pp. 5-20. Academic Press, London - New York - San Francisco 1979.

 5301 - BARDEN, J.A., FERREE, D.C.: Rootstock does not affect net photosynthesis, dark
        respiration, specific leaf weight, and transpiration of apple leaves. - J.
        Amer. Soc. hort. Sci. 104: 526-528, 1979.

 5302 - BARLOW, H.W.B.: Sectorial patterns in leaves on fruit tree shoots produced by
        radioactive assimilates and solutions. - Ann. Bot. 43: 593-602, 1979.

 5303 - BAR-NUN, N., POLJAKOFF-MAYBER, A.: Invarietal differences in the amino acid
        composition of pea roots as related to their response to salinity. - Ann. Bot.
        44: 309-314, 1979.

 5304 - BARON, W.M.M.: Organization in Plants. - Edward Arnold, London 1979.

 5305 - BARTLEY, M., HALLAM, N.D.: Changes in the fine structure of the desiccation-
        -tolerant sedge *Coleochloa setifera* (Ridley) Gilly under water stress. - Aust.
        J. Bot. 27: 531-545, 1979.

 5306 - BAR-YOSEF, B., LAMBERT, J.R.: Corn and cotton root growth in response to osmotic
        potential and oxygen and nitrate concentrations in nutrient solutions. - In:
        HARLEY, J.L., SCOTT RUSSELL, R. (ed.): The Soil-Root Interface. Pp. 287-299.
        Academic Press, London - New York - San Francisco 1979.

 5307 - BASSIRI, A., NAHAPETIAN, A.: Influences of irrigation regimens on phytate and
        mineral contents of wheat grain and estimates of genetic parameters. - J. Agr.
        Food. Chem. 27: 984-989, 1979.

*5308 - BASU, R.N., DASGUPTA, M.: Control of seed deterioration by free radical control-
        ling agents. - Ind. J. exp. Biol. 16: 1070-1073, 1978.

 5309 - BATCHELOR, J.T., SCOTT, H.D.: N and K contents of irrigated and nonirrigated
        soybean plant parts. - Arkansas Farm Res. 28(4): 6, 1979.

 5310 - BAUER, H.: Photosynthesis of ivy leaves (*Hedera helix* L.) after heat stress.
        III. Stomatal behaviour. - Z. Pflanzenphysiol. 92: 277-283, 1979.

 5311 - BEADLE, C.L., JARVIS, P.G., NEILSON, R.E.: Leaf conductance as related to xylem
        water potential and carbon dioxide concentration in Sitka spruce. - Physiol.
        Plant. 45: 158-166, 1979.

*5312 - BEAUMONT, G., BASTIN, R., THERRIEN, H.P.: Effets physiologiques de l'atrazine
        à doses sublétales sur *Lemna minor* L. I. Influence sur la croissance, la teneur
        en chlorophylle en protéines et en azote soluble et total. - Natur. Can. 103:
        527-533, 1976.

 5313 - BECKERSON, D.W., HOFSTRA, G.: Stomatal responses of white bean to O_3 and SO_2
        singly or in combination. - Atmos. Environ. 13: 533-535, 1979.

 5314 - BECKERSON, D.W., HOFSTRA, G.: Response of leaf diffusive resistance of radish,
        cucumber and soybean to O_3 and SO_2 singly or in combination. - Atmos. Environ.
        13: 1263-1268, 1979.

 5315 - BECKERSON, D.W., HOFSTRA, G., WUKASCH, R.: The relative sensitivities of 33
        bean cultivars to ozone and sulfur dioxide singly or in combination in control-
        led exposures and to oxidants in the field. - Plant Dis. Rep. 63: 478-482,
        1979.

*5316 - BEN-ASHER, J., LOMEN, D.O., WARRICK, A.W.: Linear and nonlinear models of in-
         filtration from a point source. - Soil Sci. Soc. Amer. J. 42: 3-6, 1978.

 5317 - BENGTSON, C., FALK, S.O., LARSSON, S.: Effects of kinetin on transpiration
        rate and abscisic acid content of water stressed young wheat plants. - Physiol.
        Plant. 45: 183-188, 1979.

 5318 - BENNERT, W.H., MOONEY, H.A.: The water relations of some desert plants in
        Death Valley, California. - Flora 168: 405-427, 1979.

 5319 - BENTON, A.R., Jr.: Reply to discussion by Sherwood B. Idso "Evapotranspiration
        from water hyacinth (*Eichhornia crassipes* (Mart.) Solms) in Texas Reservoirs".
        - Water Resour. Bull. 15: 1470-1472, 1979.

*5320 - BERGER, A.: L'alimentation en eau en milieu salé. - Bull. Soc. bot. France 125
        (Actualités bot. 3/4): 159-176, 1978.

 5321 - BERINGER, H., TROLLDENIER, G.: Influence of K nutrition on the response to
        environmental stress. - In: Potassium Research - Review and Trends. Pp. 189-
        222. International Potash Research Institute, Bern 1979.

*5322 - BERNER, T.: Ecophysiological activity of hypolithic desert algae. - J. Phycol.
        14 (Suppl.): 39, 1978.

 5323 - BESSIS, J., GUYOT, M.: An attempt to use stomatal characters in systematic and
        phylogenetic studies of the *Solanaceae*. - In: HAWKES, J.G., LESTER, R.N.,
        SKELDING, A.D. (ed.): Biology and Taxonomy of the *Solanaceae*. Pp. 321-326.
        Academic Press, London - New York - San Francisco 1979.

 5324 - BEWLEY, J.D.: Physiological aspects of desiccation tolerance. - Annu. Rev. Plant
        Physiol. 30: 195-238, 1979.

*5325 - BHAN, S.: Root development and moisture use of rainfed brown sarson as influenced
        by agronomic practices. - Ind. J. Agron. 21: 245-248, 1976.

*5326 - BHAN, S.: Studies on suitable agronomical practices for rainfed brown sarson in
        Uttar Pradesh. - Ind. J. Agron. 21: 271-276, 1976.

*5327 - BHATIA, D.S., MALIK, C.P.: Changing pattern of enzymes in the epidermal peelings
        with opened and closed stomata. - Biochem. Physiol. Pflanz. 172: 173-176, 1978.

 5328 - BIDWELL, R.G.S.: Plant Physiology. Second Edition. - Collier Macmillan Publishers,
        London 1979.

*5329 - BIGOT, J., BINET, P.: Modifications de l'activité pectinestérase sous l'action
        du sel. Relations avec la succulence. - Bull. Soc. bot. France 125 (Actualités
        bot. 3/4): 229-238, 1978.

*5330 - BILLARD, J.P., BOUCAUD, J.: Étude de la glutamate deshydrogénase d'un helo-
        phyte obligatoire: *Suaeda maritima* (L.) Dum. var. *macrocarpa* Moq. Effet du
        NaCl *in vivo* et *in vitro*. - Bull. Soc. bot. France 125 (Actualités bot. 3/4):
        249-258, 1978.

*5331 - BINET, P.: Introduction: Définitions et variabilités de l'halophilie et de la
        résistance aux sels. - Bull. Soc. bot. France 125 (Actualités bot. 3/4): 9-21,
        1978.

*5332 - BINET, P.: Introduction: Caractéristiques physiologiques liées à l'halophilie
        et à la résistance aux sels. - Bull. Soc. bot France 125 (Actualités bot. 3/4):
        73-93, 1978.

 5333 - BISSON, M.A., KIRST, G.O.: Osmotic adaptation in the marine alga *Griffithsia
        monilis* (*Rhodophyceae*): The role of ions and organic compounds. - Aust. J.
        Plant Physiol. 6: 523-538, 1979.

5334 - BITTMAN, S., STEPPLER, H.A.: A gasometric apparatus for monitoring evaporation rate from plant tissues during transpiration and drying. - Can. J. Plant Sci. 59: 545-548, 1979.

*5335 - BJÖRKMAN, O., BADGER, M., ARMOND, P.A.: Thermal acclimation of photosynthesis: Effect of growth temperature on photosynthetic characteristics and components of the photosynthetic apparatus in *Nerium oleander*. - Carnegie Inst. Year Book 77: 262-276, 1978.

5336 - BLACK, C.R.: The relationship between transpiration rate, water potential, and resistances to water movement in sunflower (*Helianthus annuus* L.). - J. exp. Bot. 30: 235-243, 1979.

5337 - BLACK, C.R.: The relative magnitude of the partial resistances to transpirational water movement in sunflower (*Helianthus annuus* L.). - J. exp. Bot. 30: 245-253, 1979.

5338 - BLACK, C.R.: A quantitative study of the resistances to transpirational water movement in sunflower (*Helianthus annuus* L.). - J. exp. Bot. 30: 947-953, 1979.

5339 - BLACK, C.R., BLACK, V.J.: The effects of low concentrations of sulphur dioxide on stomatal conductance and epidermal cell survival in field bean (*Vicia faba* L.). - J. exp. Bot. 30: 291-298, 1979.

5340 - BLACK, C.R., BLACK, V.J.: Light and scanning electron microscopy of SO_2 - induced injury to leaf surfaces of field bean (*Vicia faba* L). - Plant Cell Environ. 2: 61-65, 1979.

5341 - BLACK, C.R., SQUIRE, G.R.: Effects of atmospheric saturation deficit on the stomatal conductance of pearl millet (*Pennisetum typhoides* S. and H.) and groundnut (*Arachis hypogaea* L.). - J. exp. Bot. 30: 935-945, 1979.

5342 - BLACK, T.A.: Evapotranspiration from Douglas fir stands exposed to soil water deficits. - Water Resour. Res. 15: 164-170, 1979.

5343 - BLACK, V.J., UNSWORTH, M.H.: A system for measuring effects of sulphur dioxide on gas exchange of plants. - J. exp. Bot. 30: 81-88, 1979.

5344 - BLACK, V.J., UNSWORTH, M.H.: Resistance analysis of sulphur dioxide fluxes to *Vicia faba*. - Nature 282: 68-69, 1979.

*5345 - BLAD, B.L., STEADMAN, J.R., WEISS, A.: Canopy structure and irrigation influence white mold disease and microclimate of dry edible beans. - Phytopathology 68: 1431-1437, 1978.

5346 - BLAIR, G.J., MOMUAT, E.O., MAMARIL, C.P.: Sulfur nutrition of rice. II. Effect of source and rate of S on growth and yield under flooded conditions. - Agron. J. 71: 477-480, 1979.

*5347 - BLAKE, J.R.: On the hydrodynamics of plasmodesmata. - J. theor. Biol. 74: 33-47, 1978.

5348 - BLANCHET, R., GELFI, N.: Influence de réductions de la surface foliaire sur la croissance, le développement et la production d'un Soja de type indéterminé (*Glycine max*. L. Merril, cv. Amsoy 71). - C.R. Acad. Sci. Paris, Sér. D. 289: 299-302, 1979.

*5349 - BLOEMEN, G.W.: A high-accuracy recording pan-evaporimeter and some of its possibilities. - J. Hydrol. 39: 159-173, 1978.

5350 - BLOOM, A.J.: Salt requirement for Crassulacean acid metabolism in the annual succulent, *Mesembryanthemum crystallinum*. - Plant Physiol. 63: 749-753, 1979.

5351 - BLOOM, A.J.: Diurnal ion fluctuations in the mesophyll tissue of the Crassulacean acid metabolism plant *Mesembryanthemum crystallinum*. - Plant Physiol. 64:
919-923, 1979.

5352 - BLOOM, A.J., TROUGHTON, J.H.: High productivity and photosynthetic flexibility
in a CAM plant. - Oecologia 38: 35-43, 1979.

*5353 - BLUNT, C.G., HAYDOCK, K.P.: Effect of irrigation, nitrogen and defoliation on
pangola grass in the dry season at the Ord Valley, north-western Australia. -
Aust. J. exp. Agr. anim. Husb. 18: 825-833, 1978.

*5354 - BOELS, D., VAN GILS, J.B.H.M., VEERMAN, G.J., WIT, K.E.: Theory and system of
automatic determination of soil moisture characteristics and unsaturated hydraulic conductivities. - Soil Sci. 126: 191-199, 1978.

*5355 - BOÏKOV, S.: Vliyanie na polivniya rezhim i nachina na napoyavane v"rkhu s"sta-
va i khranitelnata stoĭnost na lyutsernata. [Composition and feeding value of
lucerne as influenced by the regime and method of irrigation.] - Rasteniev"dni
Nauki 15(9-10): 89-96, 1978. [In Bulg, ab: R,E.]

*5356 - BORCHERT, R.: Feedback control and age-related changes of shoot growth in
seasonal and nonseasonal climates. - In: TOMLINSON, P.B., ZIMMERMANN, M.H.
(ed.): Tropical Trees as Living Systems. Pp. 497-515. Cambridge University
Press, Cambridge - London - New York - Melbourne 1978.

*5357 - BORKOWSKA, B., SINGHA, S., KHAN, A.A.: Changes in ABA content of lettuce seeds
by osmotic, fusicoccin and other treatments. - Plant Physiol. 61(Suppl.): 94,
1978.

5358 - BÖTTCHER, H., FRÖHLICH, H., HÜBNER, C.: Ergebnisse zum komplexen Einfluss von
Beregnung, Pflanzenbestand und Düngung auf den Ertrag, die Qualität und Lager-
fähigkeit von Speisezwiebeln (*Allium cepa* L.). 1. Einfluss auf Ertrag und
Qualität. - Arch. Gartenbau 27: 283-306, 1979.

*5359 - BOUCAUD, J.: Action du NaCl sur la nutrition azotée d'un halophyte: *Suaeda
maritima* (L.) var. *macrocarpa* Moq. Emploi de l'azote de masse 15. - Bull. Soc.
bot. France 125 (Actualités bot. 3/4): 239-248, 1978.

*5360 - BOUCAUD, J., UNGAR, I.A.: Halophilie et résistance au sel dans le genre
Suaeda Forsk. - Bull. Soc. bot. France 125 (Actualités bot. 3/4): 23-35, 1978.

*5361 - BRADBEER, P.A., THOMASON, B., BRADBEER, P.: Problems of terminology in the
teaching of plant water relations. - J. Biol. Educ. 10: 299-302, 1976.

5362 - BRADFIELD, E.G., GUTTRIDGE, C.G.: The dependence of calcium transport and leaf
tipburn in strawberry on relative humidity and nutrient solution concentration.
- Ann. Bot. 43: 363-372, 1979.

*5363 - BRANDHAM, P.E., CUTLER, D.F.: Influence of chromosome variation on the organi-
sation of the leaf epidermis in a hybrid *Aloë* (*Liliaceae*). - Bot. J. Linn. Soc.
77: 1-16, 1978.

5364 - BRANDT, D.C., KÖRNER, F.J.S.M.: A small-scale saturator for experiments in
which a high accuracy in humidity is required. - Oecologia 40: 325-329, 1979.

5365 - BRAUNE, W., LEMAN, A., TAUBERT, H.: Pflanzenanatomisches Praktikum I. - VEB
Gustav Fischer Verlag, Jena 1979.

5366 - BRETERNITZ, R., ASMUS, F., GÖRLITZ, H.: Wirkung der Gülleverregnung in einer
Futterbaufruchtfolge auf Ertrag, N-Entzug und Humusgehalt einer Sand-Rosterde.
- Arch. Acker- Pflanzenbau Bodenk. 23: 555-562, 1979.

*5367 - BRETT, D.W.: Dendroclimatology of elm in London. - Tree-ring Bull. 38: 35-44,
1978.

5368 - BRETT, D.W.: Ontogeny and classification of the stomatal complex of *Platanus* L. - Ann. Bot. 44: 249-251, 1979.

*5369 - BRIENS, M.: Biogenèse de l'acide oxalique chez *Suaeda macrocarpa* Moq. - Bull. Soc. bot. France 125 (Actualités bot. 3/4): 215-221, 1978.

5370 - BRINCKMANN, E., WILLERT, D.J., von : Response of acid metabolism in CAM plants to environmental variables. - Naturwissenschaften 66: 526-527, 1979.

*5371 - BRINKHUIS, B.H., TEMPEL, N.: Photosynthesis in exposed saltmarsh macroalgae. - J. Phycol. 11(Suppl.): 11, 1975.

*5372 - BRINKMAN, M.A., FREY, K.J.: Flag leaf physiological analysis of oat isolines that differ in grain yield from their recurrent parents. - Crop.Sci. 18: 69-73, 1978.

5373 - BRIX, H.: Effects of plant water stress on photosynthesis and survival of four conifers. - Can. J. Forest Res. 9: 160-165, 1979.

*5374 - BROOKES, P.C.: Control of evaporation and surface-accumulation of nutrients in sand-culture experiments with oak seedlings. - J. appl. Ecol. 15: 635-637, 1978.

5375 - BROOKES, P.C., GILDON, A., LANE, P.W., JOHNSTON, A.E.: Effects of different watering methods, soil weights, soil diluents and soil coverings on the yield and nutrient uptake by ryegrass grown in a controlled environment. - J. Sci. Food Agr. 30: 528-531, 1979.

5376 - BROWN, A.D.: Physiological problems of water stress. - In: SHILO, M.(ed.): Strategies of Microbial Life in Extreme Environments. Pp. 65-81. Verlag Chemie, Weinheim - New York 1979.

5377 - BROWN, D.C.W., LEUNG, D.W.M., THORPE, T.A.: Osmotic requirement for shoot formation in tobacco callus. - Physiol. Plant. 46: 36-41, 1979.

5378 - BROWN, D.H., BUCK, G.W.: Desiccation effects and cation distribution in bryophytes. - New Phytol. 82: 115-125, 1979.

5379 - BROWN, F.A., Jr.: Dynamic biomagnetism associates bean seeds. - Experientia 35: 468-469, 1979.

*5380 - BROWN, K.W.: Sugar beet and potatoes. - In: MONTEITH, J.L. (ed.): Vegetation and the Atmosphere. Volume 2. Case Studies. Pp. 65-86. Academic Press, London - New York - San Francisco 1976.

5381 - BROWN, R.H., SIMMONS, R.E.: Photosynthesis of grass species differing in CO_2 fixation pathways. I. Water-use efficiency. - Crop Sci. 19: 375-379, 1979.

5382 - BRUINSMA, J.: Root hormones and overground development. - In: SCOTT, T.K. (ed.): Plant Regulation and World Agriculture. Pp. 35-47. Plenum Press, New York - London 1979.

*5383 - BRUN, W.A.: Assimilation. - In: NORMAN, A.G. (ed.): Soybean Physiology, Agronomy, and Utilization. Pp. 45-76. Academic Press, New York - London - San Francisco 1978.

5384 - BUCK, G.W., BROWN, D.H.: The effect of desiccation on cation location in lichens. - Ann. Bot. 44: 265-277, 1979.

5385 - BUCKNER, R.C., BUSH, L.P., BURRUS, P.B., II.: Succulence as a selection criterion for improved forage quality in *Lolium-Festuca* hybrids. - Crop Sci. 19: 93-96, 1979.

*5386 - BUICULESCU, I., POPESCU, D., IORDAN, M., PEICEA, I.M., ŞERBĂNESCU, G.: The influence of air-polluting gases on some plant metabolism. - Rev. Roum. Biol., Ser. Biol. vég. 23: 187-193, 1978.

*5387 - BUKHAR, I.E., MEDVEDEVA, T.N., ZHEKU, R.I.: Soderzhanie khlorofilla, sukhikh veshchestv i ovodnenost' list'ev ozimoĭ pshenitsy v zavisimosti ot usloviĭ pitaniya. [Contents of chlorophyll, dry matter and water in winter wheat leaves in dependence on nutrition.] - Izv. Akad. Nauk. Mold. SSR, Ser. biol. khim. Nauk 1975(5): 84-89, 94, 1975. [In R.]

 5388 - BUNCE, J.A., CHABOT, B.F., MILLER, L.N.: Role of annual leaf carbon balance in the distribution of plant species along an elevational gradient. - Bot. Gaz. 140: 288-294, 1979.

 5389 - BURCH, G.J.: Soil and plant resistances to water absorption by plant root systems. - Aust. J. agr. Res. 30: 279-292, 1979.

 5390 - BURKE, M.J.: Discussion: Water in plants: The phenomenon of frost survival. - In: Comparative Mechanisms of Cold Adaptation. Pp. 259-281. Academic Press, London - New York - San Francisco 1979.

*5391 - BUSBY, J.R., BLISS, L.C., HAMILTON, C.D.: Microclimate control of growth rates and habitats of the boreal forest mosses, *Tomenthypnum nitens* and *Hylocomium splendens*. - Ecol. Monogr. 48: 95-110, 1978.

*5392 - BYSTRZEJEWSKA, G.: Wpływ deficytu fosforu na fotosyntetyczna funkcje roślin. [Effect of phosphorus deficiency on plant photosynthetic functions.] - Postepy Nauk roln. 25(5): 43-54, 1978. [In Pol.]

*5393 - CAHAYLA-WYNNE, R., GLENN-LEWIN, D.C.: The forest vegetation of the Driftless Area, Northeast Iowa. - Amer. Midl. Natur. 100: 307-319, 1978.

 5394 - CALDWELL, M.M.: Root structure: The considerable cost of belowground function. - In: SOLBRIG, O.T., JAIN, S., JOHNSON, G.B., RAVEN, P.H. (ed.): Topics in Plant Population Biology. Pp. 409-432, Columbia University Press, New York 1979.

*5395 - CALDWELL, M.M., JOHNSON, D.A., FAREED, M.: Constrains on tundra productivity: Photosynthetic capacity in relation to solar radiation utilization and water stress in arctic and alpine tundras. - In: TIESZEN, L.L. (ed.): Vegetation and Production Ecology of Alaskan Arctic Tundra. Pp. 323-342. Springer-Verlag, New York - Heidelberg - Berlin 1978.

 5396 - CALE, W.G., Jr.: Modeling grassland primary productivity using piecewise stationary, piecewise linear mathematics. - Ecol. Model. 7: 107-123, 1979.

*5397 - CALLAGHAN, T.V., COLLINS, N.J., CALLAGHAN, C.H.: Photosynthesis, growth and reproduction of *Hylocomium splendens* and *Polytrichum commune* in Swedish Lapland. - Oikos 31: 73-88, 1978.

 5398 - CAMPBELL, C.A., DAVIDSON, H.R.: Effect of temperature, nitrogen fertilization and moisture stress on growth, assimilate distribution and moisture use by Manitou spring wheat. - Can. J. Plant Sci. 59: 603-626, 1979.

 5399 - CAMPBELL, C.A., DAVIDSON, H.R.: Effect of temperature, nitrogen fertilization and moisture stress on yield, yield components, protein content and moisture use efficiency of Manitou spring wheat. - Can. J. Plant Sci. 59: 963-974, 1979.

 5400 - CAMPBELL, G.S., PAPENDICK, R.I., RABIE, E., SHAYO-NGOWI, A.J.: A comparison of osmotic potential, elastic modulus, and apoplastic water in leaves of dryland winter wheat. - Agron. J. 71: 31-36, 1979.

*5401 - CANNELL, M.G.R., BRIDGWATER, F.E., GREENWOOD, M.S.: Seedling growth rates,
        water stress responses and root-shoot relationships related to eight-year
        volumes among families of *Pinus taeda* L. - Silvae Genet. 27: 237-248, 1978.

5402 - CANNELL, R.Q., GALES, K., SNAYDON, R.W., SUHAIL, B.A.: Effects of short-term
       waterlogging on the growth and yield of peas (*Pisum sativum*). - Ann. appl.
       Biol. 93: 327-335, 1979.

5403 - CARBON, B.A., BARTLE, G.A., MURRAY, A.M.: A method for visual estimation of
       leaf area. - Forest Sci. 25: 53-58, 1979.

5404 - CARBONNEAU, A., CASTERAN, P.: Irrigation-depressing effect on floral initiation
       of Cabernet Sauvignon grapevines in Bordeaux area. - Amer. J. Enol. Viticult.
       30: 3-7, 1979.

*5405 - CARDE, J.-P.: Ultrastructural studies of *Pinus pinaster* needles: the endoder-
        mis. - Amer. J. Bot. 65: 1041-1054, 1978.

5406 - CARLSON, R.E., MOMEN, N.N., ARJMAND, O., SHAW, R.H.: Leaf conductance and
       leaf-water potential relationships for two soybean cultivars grown under con-
       trolled irrigation. - Agron. J. 71: 321-325, 1979.

5407 - CARPITA, N.C., NABORS, M.W., ROSS, C.W., PETRETIC, N.L.: The growth physics
       and water relations of red-light-induced germination in lettuce seeds. III.
       Changes in the osmotic and pressure potential in the embryonic axes of red-
       and far-red-treated seeds. - Planta 144: 217-224, 1979.

5408 - CARPITA, N.C., NABORS, M.W., ROSS, C.W., PETRETIC, N.L.: The growth physics
       and water relations of red-light-induced germination in lettuce seeds. IV.
       Biochemical changes in the embryonic axes of red- and far-red-treated seeds. -
       Planta 144: 225-233, 1979.

5409 - CARR, S.G.M., CARR, D.J.: An unusual feature of stomatal microanatomy in cer-
       tain taxonomically-related *Eucalyptus* spp. - Ann. Bot. 44: 239-243, 1979.

5410 - CATHEY, G.W.: Acceleration of boll dehiscence with desiccant chemicals. -
       Agron. J. 71: 505-508, 1979.

5411 - CERDA, A., BINGHAM, F.T., HOFFMAN, G.J., HUSZAR, C.K.: Leaf water potential
       and gaseous exchange of wheat and tomato as affected by NaCl and P levels in
       the root medium. - Agron. J. 71: 27-31, 1979.

5412 - CEULEMANS, R., IMPENS, I.: Study of CO_2 exchange processes, resistances to
       carbon dioxide and chlorophyll content during leaf ontogenesis in poplar. -
       Biol. Plant. 21: 302-306, 1979.

5413 - CEULEMANS, R., IMPENS, I., GABRIËLS, R.: Comparative study of leaf water po-
       tential, diffusion resistance, and transpiration of azalea cultivars subjected
       to water stress. - HortScience 14: 507-509, 1979.

*5414 - CEULEMANS, R., LEMEUR, R., MOERMANS, R., SAMSUDDIN, Z., IMPENS, I.: Studie van
        bladanatomie en -morfologie en hun samenhang met de diffusiekarakteristieken
        van verschillende *Populus*-klonen. [Study of leaf anatomy and morphology and
        their relation with the diffusion characteristics of several *Populus* clones.]
        - Universitaire Instelling Antwerpen Rapport Nr. 2. - 1976: 1-41, 1976.
        [In Flem.]

5415 - CHAKRABARTI, D.K., BASUCHAUDHARY, K.C.: Effect of the metabolites of *Fusarium
       oxysporum* f. sp. *carthami* on transpiration of safflower. - Ind. J. Plant Phy-
       siol. 22: 116-120, 1979.

*5416 - CHAKRAVARTI, N.V.K., SASTRY, P.S.N.: Seasonal variation of pan to potential
        evapotranspiration in a semi-arid region. - Ind. J. Power & Valley Dev. 28:
        79-81, 1978.

5417 - CHANDLER, L.D., ARCHER, T.L., WARD, C.R., LYLE, W.M.: Influences of irrigation practices on spider mite densities on field corn. - Environ. Entomol. 8: 196-201, 1979.

5418 - CHARLES-EDWARDS, D.A.: A model for leaf growth. - Ann. Bot. 44: 523-535, 1979.

*5419 - CHAROLAIS, N.: Influence du sel sur la photorespiration chez le cotonnier. - Bull. Soc. bot. France 125 (Actualités bot. 3/4): 199-214, 1978.

*5420 - CHAROY, J., FOREST, F., LEGOUPIL, J.-C., BASSEREAU, D.: Besoins en eau de la canne a sucre. - Agron. Trop. 33: 344-369, 1978.

5421 - CHASE, R.L., APPLEBY, A.P.: Effect of intervals between application and tillage on glyphosate control of *Cyperus rotundus* L. - Weed Res. 19: 207-211, 1979.

*5422 - CHAUDHARY, T.N.: Rice production in relation to water management and environmental conditions. - Ind. J. Genet. Plant Breed. 39A: 55-59, 1978.

*5423 - CHAZDON, R.L.: Ecological aspects of the distribution of C_4 grasses in selected habitats of Costa Rica. - Biotropica 10: 265-269, 1978.

*5424 - CHEEMA, S.S., MALHOTRA, O.P., KUNDRA, H., SINGH, M.: Timing of last irrigation to wheat in relation to profile-charging levels. - Ind. J. agr. Sci. 48: 100-102, 1978.

*5425 - CHEN, H.-H., LI, P.H.: Interactions of low temperature, water stress, and short days in the induction of stem frost hardiness in red osier dogwood. - Plant Physiol. 62: 833-835, 1978.

*5426 - CHENG, Y.H., SCHENCK, N.C.: Effect of soil temperature and moisture on survival of the soybean root rot fungi *Neocosmospora vasinfecta* and *Fusarium solani* in soil. - Plant Dis. Rep. 62: 945-949, 1978.

*5427 - CHONAN, N.: A comparative anatomy of mesophyll among the leaves of gramineous crops. - JARQ 12: 128-131, 1978.

5428 - CHONG, C., DESJARDINS, R.L., LIN, C.S.: Changes in water and carbohydrate status of three nursery species in containers during overwinter storage in three environments. - Can. J. Plant Sci. 59: 747-755, 1979.

5429 - CHRISTIANSEN, M.N.: Physiological bases for resistance to chilling. - HortScience 14: 583-586, 1979.

5430 - CHU, A.C.P., McPHERSON, H.G., HALLIGAN, G.: Recovery growth following water deficits of different duration in prairie grass. - Aust. J. Plant Physiol. 6: 255-263, 1979.

*5431 - CLARKE, J.M., SIMPSON, G.M.: Growth analysis of *Brassica napus* cv. Tower. - Can. J. Plant Sci. 58: 587-595, 1978.

5432 - CLOUGH, J.M., ALBERTE, R.S., TEERI, J.A.: Photosynthetic adaptation of *Solanum dulcamara* L. to sun and shade environments. II. Physiological characterization of phenotypic response to environment. - Plant Physiol. 64: 25-30, 1979.

5433 - CLOUGH, J.M., TEERI, J.A., ALBERTE, R.S.: Photosynthetic adaptation of *Solanum dulcamara* L. to sun and shade environments. I. A comparison of sun and shade populations. - Oecologia 38: 13-21, 1979.

5434 - CLUCAS, R.D., LADIGES, P.Y.: Variations in populations of *Eucalyptus ovata* Labill., and the effects of waterlogging on seedling growth. - Aust. J. Bot. 27: 301-315, 1979.

5435 - COCKBURN, W., TING, I.P., STERNBERG, L.O.: Relationships between stomatal behavior and internal carbon dioxide concentration in Crassulacean acid metabolism plants. - Plant Physiol. 63: 1029-1032, 1979.

5436 - COLMAN, B., MANSON, B.T., ESPIE, G.S.: The rapid isolation of photosynthetically active mesophyll cells from *Asparagus* cladophylls. - Can. J. Bot. 57:
1505-1510, 1979.

5437 - CONESA, A.P., METTAUER, H., HAEFLINGER, R., TRENDEL, R., TUAL, Y., GROSS, P.:
Etude de la productivité de l'agrosystème betteravier en Alsace. Essai d'éta-
blissement d'un modèle empirique prédictif. - Ann. agron. 30: 281-303, 1979.

5438 - CONTOUR-ANSEL, D., LOUGUET, P.: Comparaison entre les teneurs en acides orga-
niques d'épidermes foliaires isolés de *Pelargonium X hortorum*, en relation
avec l'état d'ouverture ou de fermeture des stomates: l'acide malique joue-t-il
toujours un rôle déterminant dans les mouvements stomatiques? - Physiol. vég.
17: 337-346, 1979.

5439 - COOPER, C.S.: Yields of irrigated grass and legume pasture mixtures in the
northern Rocky Mountain area. - Agron. J. 71: 885-888, 1979.

5440 - COOPER, J.L.: Growth and yield of a semi-dwarf and a standard height wheat
cultivar in the Murrumbidgee irrigation area. - Aust. J. exp. Agr. anim. Husb.
19: 554-558, 1979.

5441 - COOPER, S.D., COCKBURN, W.: Osmotically induced water stress, potassium uptake,
and stomatal aperture in epidermal strips of *Vicia faba* L. - J. exp. Bot. 30:
913-918, 1979.

5442 - CORK, R.J., NELMES, B.J.: Vesicles in guard-cell walls and their possible roles
in the stomatal mechanism. - J. Cell Sci. 38: 83-95, 1979.

5443 - COUMANS, M., CEULEMANS, E., GASPAR, T.: Stabilized dormancy in sugarbeet fruits.
III. Water sensitivity. - Bot. Gaz. 140: 389-392, 1979.

5444 - COWAN, M.C.: Water use and phosphorus and potassium status of wheat seedlings
colonized by *Gaeumannomyces graminis* or *Phialophora radicicola*. - Plant Soil
52: 1-8, 1979.

5445 - COWGILL, U.M.: Variations in annual precipitation and selenium accumulation by
milk-vetch. - J. Plant Nutr. 1: 73-80, 1979.

*5446 - COYNE, P.I., BINGHAM, G.E.: Photosynthesis and stomatal light responses in
snap beans exposed to hydrogen sulfide and ozone. - J. Air Pollut. Cont. Assoc.
28: 1119-1123, 1978.

*5447 - CRAUBNER, H.: Hydration and partial specific volume of stroma-freed chloro-
plasts. - Colloid Polymer Sci. 253: 713-719, 1975.

5448 - CREMER, K.W., SVENSSON, J.G.P.: Changes in length of *Pinus radiata* shoots
reflecting loss and uptake of water through foliage and bark surfaces. -
Aust. Forest Res. 9: 163-172, 1979.

5449 - CRITTENDEN, P.D., KERSHAW, K.A.: Studies on lichen-dominated systems. XXII.
The environmental control of nitrogenase activity in *Stereocaulon paschale* in
spruce-lichen woodland. - Can. J. Bot. 57: 236-254, 1979.

*5450 - CROWE, J.H., CLEGG, J.S. (ed): Dry Biological Systems. - Academic Press, New
York - San Francisco - London 1978.

5451 - CRUIZIAT, P., THOMAS, D.A., BODET, C.: Comparaison entre mesures locales et
mesure globale de la résistance stomatique de feuilles de Tournesol (*Helianthus
annuus*). - Oecol. Plant. 14: 447-459, 1979.

5452 - CRUZ-ROMERO, G., RAMOS, C.: Soil water stress and air humidity effects on the
root system of sorghum. - In: HARLEY, J.L., SCOTT RUSSELL, R. (ed.): The Soil-
-Root Interface. Pp. 419-420. Academic Press, London - New York - San Francisco
1979.

5453 - CUNNINGHAM, G.L., SYVERTSEN, J.P., REYNOLDS, J.F., WILLSON, J.M.: Some effects of soil-moisture availability on above-ground production and reproductive allocation in *Larrea tridentata* (DC) Cov. - Oecologia 40: 113-123, 1979.

5454 - DANCETTE, C., HALL, A.E.: Agroclimatology applied to water management in the Sudanian and Sahelian zones of Africa. - In: HALL, A.E., CANNELL, G.H., LAWTON, H.W. (ed.): Agriculture in Semi-Arid Environments. Pp. 98-118. Springer-Verlag, Berlin - Heidelberg - New York 1979.

*5455 - DANON, A., DEGANI, H., CAPLAN, S.R.: The effect of the purple membrane of *H. halobium* on the osmotic fragility and water permeability of the cells. - In: CAPLAN, S.R., GINZBURG, M. (ed.): Energetics and Structure of Holophilic Microorganisms. Pp. 217-224. Elsevier / North-Holland Biomedical Press, Amsterdam - New York 1978.

*5456 - DANTZMAN, C.L., HODGES, E.M.: Effect of saline irrigation water on the growth of Pangola digitgrass (*Digitaria Decumbens* Stent.). - Soil Crop Sci. Soc. Florida Proc. 37: 131-134, 1978.

5457 - DAROSHKA, A.S., ASKANDARAŬ, É.É.: Dynamika fraktsyĭnaga sastavu vady ŭ listsyakh azimykh pshanits u asenni peryyad. [Dynamics of fractional composition of water in winter wheat leaves during autumn.] - Vestsi Akad. Navuk Belarus. SSR, Ser. biyal. Navuk 1979(3): 14-18, 138, 1979. [In Belorus, ab: E,R.]

5458 - DAS, V.S.R., RAGHAVENDRA, A.S.: Antitranspirants for improvement of water use efficiency of crops. - Outlook Agr. 10: 92-98, 1979.

5459 - DASBERG, S., BRESLER, E.: L'irrigation. - Recherche 10: 968-975, 1979.

5460 - Da SILVA, P.R.F., STUTTE, C.A.: Loss of gaseous N from rice leaves with transpiration. - Arkansas Farm Res. 28(4): 3, 1979.

5461 - DAULAY, H.S., SINGH, H.P., SINGH, R.P., SINGH, K.C.: Effect of different mulches on yield and moisture use of pearl millet (*Pennisetum typhoides*). - Ann. Arid Zone 18: 108-115, 1979.

5462 - DAVENPORT, D.C., ANDERSON, J.E., GAY, L.W., KYNARD, B.E., BONDE, E.K., HAGAN, R.M.: Phreatophyte evapotranspiration and its potential reduction without eradication. - Water Resour. Bull. 15: 1293-1300, 1979.

*5463 - DAVENPORT, D.C., HAGAN, R.M., GAY, L.W., KYNARD, B.E., BONDE, E.K., KREITH, F., ANDERSON, J.E.: Factors Influencing Usefulness of Antitranspirants Applied on Phreatophytes to Increase Water Supplies. - California Water Resources Center, Davis 1978.

5464 - DAVENPORT, T.L., JORDAN, W.R., MORGAN, P.W.: Movement of kinetin and gibberellic acid in leaf petioles during water stress-induced abscission in cotton. - Plant Physiol. 63: 152-155, 1979.

5465 - DAVIDSON, D.: Coleorhiza, root and coleoptile emergence and growth: Effects of different water volumes. - Can. J. Plant Sci. 59: 61-67, 1979.

*5466 - DAVIDSON, R.L.: Root systems - the forgotten component of pastures. - In: WILSON, J.R. (ed.): Plant Relations in Pastures. Pp. 86-94. CSIRO, Melbourne 1978.

5467 - DAVIES, F.S., LAKSO, A.N.: Water stress responses of apple trees. I. Effects of light and soil preconditioning treatments on tree physiology. - J. Amer. Soc. hort. Sci. 104: 392-395, 1979.

5468 - DAVIES, F.S., LAKSO, A.N.: Water stress responses of apple trees. II. Resistance and capacitance as affected by greenhouse and field conditions. - J. Amer. Soc. hort. Sci. 104: 395-397, 1979.

5469 - DAY, W.: Water stress and crop growth. - In: JOHNSON, C.B. (ed.): Physiological Processes Limiting Plant Productivity. University of Nottingham, Sutton Bonington 1979.

5470 - DAY, W., PARKINSON, K.J.: Importance to gas exchange of mass flow of air through leaves. - Plant Physiol. 64: 345-346, 1979.

5471 - DeBOER, D.W.: Comparison of three field methods for determining saturated hydraulic conductivity. - Trans. ASAE 22: 569-572, 1979.

5472 - De GREEF, J., VERBELEN, J.P., CAUBERGS, R., MOEREELS, E., SPRUYT, E.: Comparative study of photosynthetic efficiency and leaf architecture during leaf development. - In: MARCELLE, R., CLIJSTERS, H., Van POUCKE, M. (ed.): Photosynthesis and Plant Development. Pp. 49-56. Dr. W. Junk bv Publishers, The Hague - Boston - London 1979.

5473 - De JONG, R., BEST, K.F.: The effect of soil water potential, temperature and seeding depth on seedling emergence of wheat. - Can. J. Soil Sci. 59: 259-264, 1979.

5474 - De JONG, R., CAMERON, D.R.: Computer simulation model for predicting soil water content profiles. - Soil Sci. 128: 41-48, 1979.

5475 - De JONG, T.M., BARBOUR, M.G.: Contributions to the biology of *Atriplex leucophylla*, a C_4 Californian beach plant. - Bull. Torrey bot. Club 106: 9-19, 1979.

*5476 - DEKOV, D., RADKOV, P., PAVLOVA, S.: V"rkhu sukhoustoĭchivostta na nyakoi raĭonirani i perspektivni sortove fasul v"v vr"zka s"s selektsiyata. [On the drought resistance of zonated and promising bean varieties in relation to breeding.] - Rasteniev"dni Nauki 15(5): 51-58, 1978. [In Bulg, ab: R,E.]

*5477 - DEN HARTOG - VAN TER THOLEN, R.M.: Epidermal characters of the *Celastraceae* (*Hippocrateaceae* included). - Acta bot. Neerl. 27: 142, 1978.

5478 - DENNIS, R., WEBSTER, G.: Guar shows potential as drought-tolerant summer crop for Arizona grain farmers. - Progress Agr. Arizona 30: 10-11, 1979.

5479 - DEPUIT, E.J.: Photosynthesis and respiration of plants in the arid ecosystem. - In: PERRY, R.A., GOODAL, D.W. (ed.): Arid-land Ecosystems: Structure, Functioning and Management. Volume 1. Pp. 509-536. Cambridge University Press, Cambridge 1979.

*5480 - DERCO, M.: Podiel agrotechnických opatrení na prírastkoch úrod poľných plodín v závlahových podmienkach a metóda jeho stanovenia. [Share of various agrotechnical practices in yield increments of irrigated crops and method of determining this share.] - Rost. Výroba (Praha) 24: 1127-1134, 1978. [In Slov, ab: R,E,G.]

*5481 - DERCO, M.: Závlaha a kvalita cukrovej repy. [Irrigation and quality of sugar beet.] - Rost. Výroba (Praha) 24: 1203-1206, 1978. [In Slov, ab: R,E,G.]

*5482 - DERCO, M., BARTA, V.: Vplyv ošetrovania porastu cukrovej repy na úrodu a na hospodárenie s pôdnou vodou. [Influence of cultivation techniques of sugar beet crop on yield and water regime of soil.] - Rost. Výroba (Praha) 24: 1193 -1202, 1978. [In Slov, ab: R,E,G.]

5483 - DERCO, M., RAKOVAN, J.: Účinok orby a techniky hnojenia na úrodu zrna kukurice v závislosti od úrovne vodného režimu pôdy. [The effect of ploughing and fertilization technique on maize grain yields as depending on the level of water regime in the soil.] Rost. Výroba (Praha) 25: 961- 970, 1979. [In Slov, ab:R, E,G.]

5484 - DERCO, M., RAKOVAN, J.: Vplyv zálviahy a techniky hnojenia na úrodu zrna kukurice. [The effect of irrigation and fertilization method on the yields of maize
grain.] - Rost. Výroba (Praha) 25: 1267-1274, 1979. [In Slov, ab: R,E,G.]

5485 - DETLING, J.K.: Processes controlling blue grama production on the shortgrass
prairie. - In: FRENCH, N.R. (ed.): Perspectives in Grassland Ecology. Pp. 25-
42. Springer-Verlag, New York - Heidelberg - Berlin 1979.

5486 - DEY, R., VAN ALFEN, N.K.: Influence of *Corynebacterium insidiosum* on water
relations of alfalfa. - Phytopathology 69: 942-946, 1979.

*5487 - DHARMALINGAM, C., BASU, R.N.: Control of seed deterioration in cotton (*Gossypium hirsutum* L.). - Curr. Sci. 47: 484-487, 1978.

*5488 - DHINGRA, O.D.: Internally seedborne *Fusarium semitectum* and *Phomosis sp.*affecting dry and snap bean seed quality. - Plant Dis. Rep. 62: 509-512, 1978.

*5489 - DHINGRA, O.D., Da SILVA, J.F.: Effect of weed control on the internally seed-
borne fungi in soybean seeds. - Plant Dis. Rep. 62: 513-516, 1978.

5490 - DHINGRA, O.D., MUCHOVEJ, J.J.: Pot rot, seed rot, and root rot of snap bean
and dry bean caused by *Fusarium semitectum*. - Plant Dis. Rep. 63: 84-87, 1979.

*5491 - DIAMANDIS, S.: "Top-dying"of Norway spruce, *Picea abies* (L.) Karst. with special reference to *Rhizosphaera kalkhoffii* Bubák. III. Moisture content of
current and second year needles of Norway spruce. - Europ. J. Forest Pathol.
8: 357-361, 1978.

5492 - DIEST, A., van: Factors affecting the availability of potassium in soils. -
In: Potassium Research - Review and Trends. Pp. 75-97. International Potash
Research Institute, Bern 1979.

5493 - DILKS, T.J.K., PROCTOR, M.C.F.: Photosynthesis, respiration and water content
in bryophytes. - New Phytol. 82: 97-114, 1979.

5494 - DIRKSEN, C., HUBER, M.J.: Effect of irrigation frequency, and water salinity
and quantity on root water uptake distribution and leaf water potentials of
alfalfa. - In: HARLEY, J.L., SCOTT RUSSELL, R. (ed.): The Soil-Root Interface.
Pp. 421-422. Academic Press, London - New York - San Francisco 1979.

5495 - DIRKSEN, C., OSTER, J.D., RAATS, P.A.C.: Water and salt transport, water uptake, and leaf water potentials during regular and suspended high frequency irrigation of citrus. - Agr. Water Manage. 2: 241-256, 1979.

5496 - DISRAELI, D.J., FONDA, R.W.: Gradient analysis of the vegetation in a brackish
marsh in Bellingham Bay, Washington. - Can. J. Bot. 57: 465-475, 1979.

5497 - DITTRICH, P., MAYER, M., MEUSEL, M.: Proton-stimulated opening of stomata in
relation to chloride uptake by guard cells. - Planta 144: 305-309, 1979.

5498 - DODD, J.L., LAUENROTH, W.K.: Analysis of the response of a grassland ecosystem
to stress. - In: FRENCH, N.R. (ed.): Perspectives in Grassland Ecology. Pp.
43-58. Springer-Verlag, New York - Heidelberg - Berlin 1979.

5499 - DOLINER, L.H., JOLLIFFE, P.A.: Ecological evidence concerning the adaptive
significance of the C_4 dicarboxylic acid pathway of photosynthesis. - Oecologia
38: 23-34, 1979.

5500 - DONALDSON, E., NILAN, R.A., KONZAK, C.F.: The interaction of oxygen, radiation
exposure and seed water content on γ-irradiated barley seeds. - Environ. exp.
Bot. 19: 153-164, 1979.

*5501 - DONNARI, M.A., ROSELL, R.A., TORRE, L.: Productividad del ajo. II. Evapotranspi-
ración real y necesidad de agua. [Garlic productivity. II. Real evapotranspiration and water needs.] - Turrialba 28: 331-337, 1978. [In Span, ab: E.]

*5502 - DORGELO, J.: Intertidal fucoid zonation and desiccation. - Hydrobiol. Bull. (Amsterdam) 10: 115-122, 1976.

5503 - DOSTANOVA, R.Kh., KLYSHEV, L.K., TOIBAEVA, K.A.: Fenol'nye soedineniya korneĭ gorokha pri zasolenii sredy. [Phenol compounds of pea roots under salinity.] - Fiziol. Biokhim. kul't. Rast 11: 40-47, 1979. [In R, ab: E.]

5504 - DOUCE, R., JOYARD, J.: Structure and function of the plastid envelope. - Adv. bot. Res. 7: 1-116, 1979.

5505 - DOYLE, A.D., FISCHER, R.A.: Dry matter accumulation and water use relationships in wheat crops. - Aust. J. agr. Res. 30: 815-829, 1979.

*5506 - DREW, A.P., BAZZAZ, F.A.: Variations in distribution of assimilate among plant parts in three populations of *Populus deltoides*. - Silvae Genet. 27: 189-193, 1978.

5507 - DREW, A.P., BAZZAZ, F.A.: Response of stomatal resistance and photosynthesis to night temperature in *Populus deltoides*. - Oecologia 41: 89-98, 1979.

5508 - DREW, A.P., FERRELL, W.K.: Seasonal changes in the water balance of Douglas-fir (*Pseudotsuga menziesii*) seedlings grown under different light intensities. - Can. J. Bot. 57: 666-674, 1979.

5509 - DREW, M.C.: Properties of roots which influence rates of absorption. - In: HARLEY, J.L., SCOTT RUSSELL, R. (ed.): The Soil-Root Interface. Pp. 21-38. Academic Press, London - New York - San Francisco 1979.

5510 - DREW, M.C., SISWORO, E.J.: The development of waterlogging damage in young barley plants in relation to plant nutrient status and changes in soil properties. - New Phytol. 82: 301-314, 1979.

5511 - DREW, M.C., SISWORO, E.J., SAKER, L.R.: Alleviation of waterlogging damage to young barley plants by application of nitrate and a synthetic cytokinin, and comparison between the effects of waterlogging, nitrogen deficiency and root excision. - New Phytol. 82: 315-329, 1979.

5512 - DREWITT, E.G.: Effect of previous cropping on irrigation and nitrogen responses in spring-sown "Karamu" wheat. - N. Zeal. J. exp. Agr. 7: 71-78, 1979.

5513 - DREWITT, E.G.: Relationships between grain yield, grain N% and grain weight in irrigated and non-irrigated late winter- and spring-sown "Karamu" wheat. - N. Zeal. J. exp. Agr. 7: 169-173, 1979.

5514 - DUKE, S.H., SCHRADER, L.E., HENSON, C.A., SERVAITES, J.C., VOGELZANG, R.D., PENDLETON, J.W.: Low root temperature effects on soybean nitrogen metabolism and photosynthesis. - Plant Physiol. 63: 956-962, 1979.

*5515 - DUNIN, F.X., ASTON, A.R., REYENGA, W.: Evaporation from a *Themeda* grassland. II. Resistance model of plant evaporation. - J. appl. Ecol. 15: 847-858, 1978.

5516 - DUNIWAY, J.M.: Water relations of water molds. - Annu. Rev. Phytopathol. 17: 431-460, 1979.

*5517 - DURBIN, R.D., UCHYTIL, T.F.: The effect of tentoxin on fusicoccin-induced stomatal opening. - Phytopathol. Mediter. 15: 62-63, 1976.

5518 - DÜRING, H.: Wirkungen der Luft- und Bodenfeuchtigkeit auf das vegetative Wachstum und den Wasserhaushalt bei Reben. - Vitis 18: 211-220, 1979.

*5519 - DURKIN, D.: Some characteristics of flow through isolated rose stem segments. - HortScience 13: 376, 1978.

*5520 - DURKIN, D.: Effect of millipore filtration on water relations of cut rose flowers. - HortScience 13: 376, 1978.

5521 - DURKIN, D.J.: Some characteristics of water flow through isolated rose stem segments. - J. Amer. Soc. hort. Sci. 104: 777-783, 1979.

5522 - DURKIN, D.J.: Effect of millipore filtration, citric acid, and sucrose on peduncle water potential of cut rose flower. - J. Amer. Soc. hort. Sci. 104: 860-863, 1979.

5523 - DWIVEDI, S., KAR, M., MISHRA, D.: Biochemical changes in excised leaves of *Oryza sativa* subjected to water stress. - Physiol. Plant. 45: 35-40, 1979.

5524 - DWIVEDI, S., KAR, M., MISHRA, D.: Inorganic pyrophosphatase activity in water stressed excised rice leaves. - Irrig. Sci. 1: 119-124, 1979.

*5525 - DYKYJOVÁ, D., KVĚT, J. (ed.): Pond Littoral Ecosystems. Structure and Functioning. Methods and Results of Quantitative Ecosystem Research in the Czechoslovakian IBP (International Biological Program) Wetland Project. (Ecological Studies Vol. 28.). - Springer-Verlag, Berlin - Heidelberg - New York 1978.

*5526 - EAGLESON, P.S.: Climate, soil, and vegetation 1. Introduction to water balance dynamics. - Water Resour. Res. 14: 705-712, 1978.

*5527 - EAGLESON, P.S.: Climate, soil, and vegetation 2. The distribution of annual precipitation from observed storm sequences. - Water Resour. Res. 14: 713-721, 1978.

*5528 - EAGLESON, P.S.: Climate, soil, and vegetation 3. A simplified model of soil moisture movement in the liquid phase. - Water Resour. Res. 14: 722-730, 1978.

*5529 - EAGLESON, P.S.: Climate, soil, and vegetation 4. The expected value of annual evapotranspiration. - Water Resour. Res. 14: 731-739, 1978.

*5530 - EAGLESON, P.S.: Climate, soil, and vegetation 6. Dynamics of the annual water balance. - Water Resour. Res. 14: 749-764, 1978.

*5531 - EAGLESON, P.S.: Climate, soil, and vegetation 7. A derived distribution of annual water yield. - Water Resour. Res. 14: 765-776, 1978.

5532 - EAVIS, B.W., TAYLOR, H.M.: Transpiration of soybeans as related to leaf area, root length, and soil water content. - Agron. J. 71: 441-445, 1979.

5533 - ECK, H.V., MUSICK, J.T.: Plant water stress effects on irrigated grain sorghum. I. Effects on yield. - Crop Sci. 19: 589-592, 1979.

5534 - ECK, H.V., MUSICK, J.T.: Plant water stress effects on irrigated grain sorghum. II. Effects on nutrients in plant tissues. - Crop Sci. 19: 592-598, 1979.

5535 - EDWARDS, G.E., HUBER, S.C.: C4 metabolism in isolated cells and protoplasts. - In: GIBBS, M., LATZKO, E. (ed.): Photosynthesis II. Pp. 102-112. Springer-Verlag, Berlin - Heidelberg - New York 1979.

*5536 - EDWARDS, K., JOHNSTON, W.H.: Agricultural climatology of the upper Murrumbidgee River valley, New South Wales. - Aust. J. agr. Res. 29: 851-862, 1978.

5537 - EDWARDS, M., MEIDNER, H.: Direct measurements of turgor pressure potentials. IV. Naturally occurring pressures in guard cells and their relation to solute and matric potentials in the epidermis. - J. exp. Bot. 30: 829-837, 1979.

5538 - EHLERINGER, J.R.: Photosynthesis and photorespiration: Biochemistry, physiology, and ecological implications. - HortScience 14: 217-222, 1979.

*5539 - EHLERINGER, J.R., BJÖRKMAN, O.: Pubescence and leaf spectral characteristics in a desert shrub, *Encelia farinosa*. - Oecologia 36: 151-162, 1978.

*5540 - EHLERINGER, J.R., MOONEY, H.A.: Leaf hairs: effects on physiological activity and adaptive value to a desert shrub. - Oecologia 37: 183-200, 1978.

5541 - EHLERINGER, J.R., MOONEY, H.A., BERRY, J.A.: Photosynthesis and microclimate of *Camissonia claviformis*, a desert winter annual. - Ecology 60: 280-286, 1979.

5542 - EHRET, D.L., BOYER, J.S.: Potassium loss from stomatal guard cells at low water potentials. - J. exp. Bot. 30: 225-234, 1979.

5543 - EICKMEIER, W.G.: Photosynthetic recovery in the resurrection plant *Selaginella lepidophylla* after wetting. - Oecologia 39: 93-106, 1979.

5544 - EICKMEIER, W.G.: Eco-physiological differences between high and low elevation CAM species in Big Bend National Park, Texas. - Amer. Midl. Natur. 101: 118-126, 1979.

*5545 - EICKMEIER, W.G., ADAMS, M.S.: Gas exchange in *Agave lecheguilla* Torr. (*Agavaceae*) and its ecological implications. - Southwestern Natur. 23: 473-486, 1978.

5546 - ELIÁŠ, P.: Transpiračné odpory listov štyroch druhov dospelých lesných stromov merané difúznym pórometrom na prirodzenom lesnom stanovišti. [Transpirational resistances of leaves in four species of adult forest trees measured with diffusion porometer in natural forest stand.] - Folia dendrol. 1979(5): 109-142, 1979. [In Slov, ab: R,E.]

5547 - ELIÁŠ, P.: Leaf diffusion resistance pattern in an oak-hornbeam forest. - Biol. Plant. 21: 1-8, 1979.

5548 - ELIÁŠ, P.: Stomatal oscillations in adult forest trees in natural environment. - Biol. Plant. 21: 71-74, 1979.

5549 - ELIÁŠ, P.: Stomatal activity within the crowns of tall deciduous trees under forest conditions. - Biol. Plant. 21: 266-274, 1979.

5550 - ELIÁŠ, P.: Contribution to the ecophysiological study of the water relations of forest shrubs. - Preslia (Praha) 51: 77-90, 1979.

*5551 - ELIÁŠ, P., HUZULÁK, J.: Príspevok k štúdiu vodných vzťahov imelovca a cera. [Contribution to the study of water relationships between hemiparasite (*Loranthus europeus* L.) and its host (*Quercus cerris* L.).] - Acta Bot. Slov. Acad. Sci. Slov., Ser. A 4: 265-276, 1978. [In Slov, ab: R,E.]

5552 - ELKIEY, T., ORMROD, D.P.: Ozone and sulphur dioxide effects on leaf water potential of *Petunia*. - Z. Pflanzenphysiol. 91: 177-181, 1979.

5553 - ELKIEY, T., ORMROD, D.P., PELLETIER, R.L.: Stomatal and leaf surface features as related to the ozone sensitivity of *Petunia* cultivars. - J. Amer. Soc. hort. Sci. 104: 510-514, 1979.

5554 - ELKINS, C.B., HAALAND, R.L., RODRIGUEZKABANA, R., HOVELAND, C.S.: Plant-parasitic nematode effects on water use and nutrient uptake of a small-rooted and a large-rooted tall fescue genotype. - Agron. J. 71: 497-500, 1979.

5555 - ELTON, W.M., MEENTEMEYER, V.: Environmental locations of the largest trees of eastern deciduous species. - Amer. Midl. Natur. 101: 182-190, 1979.

5556 - ENGLISH, S.D., McWILLIAM, J.R., SMITH, R.C.G., DAVIDSON, J.L.: Photosynthesis and partitioning of dry matter in sunflower. - Aust. J. Plant Physiol. 6: 149-164, 1979.

5557 - ERICKSON, P.I., KIRKHAM, M.B.: Growth and water relations of wheat plants with roots split between soil and nutrient solution. - Agron. J. 71: 361-364, 1979.

5558 - ERICKSON, P.I., KIRKHAM, M.B., STONE, J.F.: Growth, water relations, and yield of wheat planted in four row directions. - Soil Sci. Soc. Amer. J.: 43: 570-574, 1979.

5559 - ERICSSON, A.: Effects of fertilization and irrigation on the seasonal changes of carbohydrate reserves in different age-classes of needle on 20-year-old Scots pine trees (*Pinus silvestris*). - Physiol. Plant. 45: 270-280, 1979.

5560 - ERLANDSSON, G.: Efflux of potassium from wheat roots induced by changes in the water potential of the root medium. - Physiol. Plant. 47: 1-6, 1979.

*5561 - ERWIN, D.C., TSAI, S.D., KHAN, R.A.: Reduced number of microsclerotia formed by *Verticillium dahliae* in cotton tissue exposed to systemic benzimidazole fungicides and desiccation. - Phytopathology 68: 1488-1494, 1978.

5562 - ESCAMILLA, E., VOSS, R., WEBB, J.R.: Effects of prices and moisture stress on nitrogen rates that maximize two economic criteria for corn. - Agron. J. 71: 609-612, 1979.

*5563 - EVANS, J.: Some growth effects of hail damage and drought in *P. patula* plantations. - South Afric. Forestry J. 105: 8-12, 1978.

*5564 - EVANS, L.T., WARDLAW, I.F.: Aspects of the comparative physiology of grain yield in cereals. - Adv. Agron. 28: 301-359, 1976.

*5565 - EYE, L.L., SNEH, B., LOCKWOOD, J.L.: Factors affecting zoospore production by *Phytophthora megasperma* var. *sojae*. - Phytopathology 68: 1766-1768, 1978.

*5566 - FADL, O.A.A.: Evapotranspiration measured by a neutron probe on Sudan Gezira vertisols. - Exp. Agr. 14: 341-347, 1978.

*5567 - FAHEY, R.C., MIKOLAJCZYK, S.D., BRODY, S.: Correlation of enzymatic activity and thermal resistance with hydration state in ungerminated *Neurospora* conidia. - J. Bacteriol. 135: 868-875, 1978.

5568 - FAHEY, T.J.: The effect of night frost on the transpiration of *Pinus concorta* ssp. *latifolia*. - Oecol. Plant. 14: 483-490, 1979.

*5569 - FAIZ, S.M.A., WEATHERLEY, P.E.: Further investigations into the location and magnitude of the hydraulic resistances in the soil:plant system. - New Phytol. 81: 19-28, 1978.

5570 - FAROOQUI, P.: Sequence of stomatal meristemoid formation in some *Leguminosae*. - Curr. Sci. 48: 489-490, 1979.

5571 - FAROOQUI, P.: Development of stomata in *Barringtonia racemosa* Roxb. - Curr. Sci. 48: 601-603, 1979.

5572 - FAROOQUI, P.: On the occurrence of abnormal stomata in plants. - Curr. Sci. 48: 841-849, 1979.

5573 - FARQUHAR, G.D.: Carbon assimilation in relation to transpiration and fluxes of ammonia. - In: MARCELLE, R., CLIJSTERS, H., Van POUCKE, M. (ed.): Photosynthesis and Plant Development. Pp. 321-328. Dr. W. Junk bv Publishers, The Hague - Boston - London 1979.

*5574 - FARQUHAR, G.D., RASCHKE, K.: On the resistance to transpiration of the sites of evaporation within the leaf. - Plant Physiol. 61: 1000-1065, 1978.

5575 - FASEHUN, F.E.: Effect of soil matric potential on leaf water potential, diffusive resistance, growth and development of *Gmelina arborea* L. seedlings. - Biol. Plant. 21: 100-104, 1979.

*5576 - FAWUSI, M.O.A.: Emergence and seedling growth of pepper as influenced by soil compaction, nutrient status and moisture regime. - Sci. Hort. 9: 329-335, 1978.

*5577 - FEDDES, R.A., KOWALIK, P.J., ZARADNY, H.: Simulation of Field Water Use and Crop Yield. - Pudoc, Wageningen 1978.

*5578 - FEDDES, R.A., ZARADNY, H.: Numerical model for transient water flow in non-homogenous soil-root systems with groundwater influence. - In: VANSTEENKISTE, (ed.): Modeling, Identification and Control in Environmental Systems. Pp. 291-307. Elsevier / North-Holland Biomedical Press, Amsterdam - New York 1978.

5579 - FEDERER, C.A.: A soil-plant-atmosphere model for transpiration and availability of soil water. - Water Resour. Res. 15: 555-562, 1979.

*5580 - FEDERER, C.A., LASH, D.: Simulated streamflow response to possible differences in transpiration among species of hardwood trees. - Water Resour. Res. 14: 1089-1097, 1978.

*5581 - FEDOSEEVA, G.P.: Fenotipicheskaya izmenchivost' mezostruktury i funktsional'-noĭ aktivnosti fotosinteticheskogo apparata. [Phenotypical variability of meso-structure and functional activity of photosynthetic tissues.] - In: Mezostruk-tura i Funktsional'naya Aktivnost' Fotosinteticheskogo Apparata. Pp. 112-131. Ural'skiĭ Gosudarstvennyĭ Universitet, Sverdlovsk 1978. [In R.]

5582 - FEDYK, Ya.D.: Lystovydni plastynky protonemy *Tetraphis pellucida* Hedw. [Leaf-like protonema plates of *Tetraphis pellucida* Hedw.] - Ukr. bot. Zh. 36: 565-569, 624, 1979. [In Ukr, ab: E,R.]

*5583 - FÉRARD, G., BINET, P.: Importance du transport des ions vers les feuilles dans la résistance au sel de *Plantago maritima* L. et de *Plantago lanceolata* L. - Bull. Soc. bot. France 125 (Actualités bot. 3/4): 105-110, 1978.

5584 - FERGUSON, H., KRALL, J.: Crop production systems in arid and semi-arid cool temperate zones. - In: THORNE, D.W., THORNE, M.D. (ed.): Soil, Water and Crop Production. Pp. 164-182. AVI Publishing Company, Inc., Westport 1979.

5585 - FERREIRA, L.G.R., DE SOUZA, J.G., PRISCO, J.T.: Effects of water deficit on proline accumulation and growth of two cotton genotypes of different drought resistances. - Z. Pflanzenphysiol. 93: 189-199, 1979.

*5586 - FERRON, F., COUDRET, A., GAUDILLÈRE, J.P.: Échanges et modes de fixation du gaz carbonique chez *Plantago maritima* L. var. *graminaea* et *Plantago lanceolata* L. sous l'action de la salinité du milieu de culture. - Bull. Soc. bot. France 125 (Actualités bot. 3/4): 189-198, 1978.

*5587 - FERWERDA, J.-D.: Oil palm. - In: ALVIM, P. de T., KOZLOWSKI, T.T. (ed.): Eco-physiology of Tropical Crops. Pp. 351-382. Academic Press, New York - San Francisco - London 1977.

5588 - FETCHER, N.: Water relations of five tropical tree species on Barro Colorado Island, Panama. - Oecologia 40: 229-233, 1979.

5589 - FIEDLER, W.: Einfluss von Bodenpflege, Bewässerung und Stickstoffdüngung auf Apfelniederstämme im VEG Stroga. - Arch. Gartenbau 27: 243-256, 1979.

*5590 - FILEV, D.S., GUĬDA, N.I., GORSHKOV, A.E.: Vlagoobespechennost' gibridov kuku-ruzy raznoĭ skorospelosti v svyazi so srokami ikh poseva. [Water supply of maize hybrids of diverse fast-ripening as connected with seeding date.] - Byul. vses. nauch.-issled. Inst. Kukuruzy 1975(4): 7-10, 1975. [In R.]

*5591 - FILIPPOV, G.L., MAKSIMOVA, L.A.: Vliyanie optimizatsii usloviĭ vyrashchivaniya na produktivnost' fotosinteza kukuruzy. [Effects of optimization of growing conditions on the net assimilation rate in maize.] - Byul. vses. nauch.-issled. Inst. Kukuruzy 1975(2): 37-40, 1975. [In R.]

*5592 - FILIPPOV, G.L., VISHNEVSKIĬ, N.V., MAKSIMOVA, L.A.: Osobennosti fotosinteti-
cheskoĬ deyatel'nosti gibridov kukuruzy v usloviyakh orosheniya. [Peculiari-
ties of photosynthetic activity of irrigated maize hybrids.] - Byull. vses.
nauch.-issled. Inst. Kukuruzy 1975(4): 19-22, 1975. [In R.]

5593 - FISCHER, R.A.: Growth and water limitation to dryland wheat yield in Australia:
a physiological framework. - J. Aust. Inst. agr. Sci. 45: 83-94, 1979.

5594 - FISCHER, R.A., SANCHEZ, M.: Drought resistance in spring wheat cultivars. II
Effects on plant water relations. - Aust. J. agr. Res. 30: 801-814, 1979.

5595 - FISCUS, E.L., MARKHART, A.H., III: Relationships between root system water
transport properties and plant size in Phaseolus. - Plant Physiol. 64: 770-
773, 1979.

5596 - FISHER, N.M., BROWNING, G.: Some effects of irrigation and plant density on
the water relations of high density coffee (Coffea arabica L.) in Kenya. -
J. hort. Sci. 54: 13-22, 1979.

*5597 - FISHER, R.F.: Juglone inhibits pine growth under certain moisture regimes. -
Soil Sci. Soc. Amer. J. 42: 801-803, 1978.

5598 - FISYUNOV, A.V., OSTAPENKO, N.A.: Regeneratsionnaya sposobnost' kornevoĬ sistemy Acroptilon repens (L.) DC. (Asteraceae) pri razlichnoĬ vlazhnosti pochvy.
[Regeneration capacity of the root system of Acroptilon repens (L.) DC.
(Asteraceae) under different soil moisture level.] - Bot. Zh. 64: 435-438,
1979. [In R.]

*5599 - FITZGERALD, P.D.: The relationship between irrigated pasture production and
radiation. - Nordic. Hydrol. 9: 263-266, 1978.

5600 - FLÜCKIGER, W., OERTLI, J.J., FLÜCKIGER, H.: Relationship between stomatal
diffusive resistance and various applied particle sizes on leaf surfaces.- Z.
Pflanzenphysiol. 91: 173-175, 1979.

5601 - FOCHT, D.D., MARTIN, J.P.: Microbiological and biochemical aspects of semi-arid
agricultural soils. - In: HALL, A.E., CANNELL, G.H., LAWTON, H.W. (ed.): Agriculture in Semi-Arid Environments. Pp. 119-147. Springer-Verlag, Berlin -
Heidelberg - New York 1979.

5602 - FOCK, H., KLUG, K., CANVIN, D.T.: Effect of carbon dioxide and temperature on
photosynthetic CO_2 uptake and photorespiratory CO_2 evolution in sunflower
leaves. - Planta 145: 219-223, 1979.

5603 - FOCK, H., LAWLOR, D.W.: Der Einfluss von Wassermangel auf den Gaswechsel und
den primären C-Stoffwechsel von Helianthus annuus und Zea mays. - Ber. Deut.
bot. Ges. 92: 145-152, 1979.

5604 - FOMISHYNA, R.M.: Vplyv raptovogo zasolennya na rozklad khlorofilu v izol'o-
vanykh lystkakh Beta vulgaris L. [Effect of sudden salting on chlorophyll
decomposition in isolated leaves of Beta vulgaris L.] - Ukr. bot. Zh. 36: 10
-13, 94, 1979. [In Ukr, ab: E,R.]

*5605 - FORD, E.D., DEANS, J.D.: The effects of canopy structure on stemflow, through-
fall and interception loss in a young Sitka spruce plantation. - J. appl. Ecol.
15: 905-917, 1978.

*5606 - FORDHAM, R.: Tea. - In: ALVIM, P.de T., KOZLOWSKI, T.T. (ed.): Ecophysiology
of Tropical Crops. Pp. 333-349. Academic Press, New York - San Francisco -
London 1977.

*5607 - FOSTER, A.C., MAUN, M.A.: Effects of highway deicing agents on Thuja occidenta-
lis in a greenhouse. - Can. J. Bot. 56: 2760-2766, 1978.

*5608 - FOUQUÉ, A., COMBRES, J.C.: Etudes sur l'irrigation en Côte d'Ivoire. - Fruits
          33: 845-847, 1978.

 5609 - FOWLER, J.L.: In-furrow water injection for improving cotton stand establish-
          ment. - Agron. J. 71: 453-458, 1979.

 5610 - FRANCIS, C.A.: Small farm cropping systems in the tropics. - In: THORNE, D.W.,
          THORNE, M.D. (ed.): Soil, Water and Crop Production. Pp. 318-348. AVI Publishing
          Company, Inc., Westport 1979.

 5611 - FRANÇOIS, J., RENARD, C.: Étude en milieu contrôlé du comportement d'un tapis
          de *Festuca arundinacea* Schreb.en régime d'assèchement. - Oecol. Plant. 14:
          417-433, 1979.

*5612 - FRANKE, G., HASSANEIN, A.E.: Zum Einfluss von GS, CCC, MH und NES auf Keimung
          und Jugendentwicklung von *Zea mays* L. bei unterschiedlicher NaCl-Substratver-
          salzung. - Beitr. trop. Landwirtsch. Veterinärmed. 14: 361-367, 1976.

*5613 - FRANQUIN, P., FOREST, F.: Un programme de simulation de l'irrigation comple-
          mentaire aux pluies. - Agron. trop. 33: 377-380, 1978.

*5614 - FREEBERG, L.R., WILSON, W.B.: Photo-oxidation and degradation of chlorophyll a
          in marine phytoplankton samples. - J. Phycol. 13(Suppl.): 22, 1977.

 5615 - FREEMAN, B., ALBRIGO, L.G., BIGGS, R.H.: Cuticular waxes of developing leaves
          and fruit of blueberry, *Vaccinium ashei* Reade cv. Bluegem. - J. Amer. Soc.
          hort. Sci. 104: 398-403, 1979.

 5616 - FREEMAN, B.M., LEE, T.H., TURKINGTON, C.R.: Interaction of irrigation and pruning
          level on growth and yield of Shiraz vines. - Amer. J. Enol. Viticult. 30: 218-
          223, 1979.

 5617 - FRENCH, N.R. (ed.): Perspectives in Grassland Ecology. (Ecological Studies 32).
          - Springer-Verlag, New York - Heidelberg - Berlin 1979.

*5618 - FRENYÖ, V.: Gleditsia-magvak vízfelvétele és légzésmenete. [Respiration rate
          of *Gleditsia* seeds in dependence on water uptake.] - Bot. Közlem. 65: 209-214,
          1978. [In Hung, ab: G.]

 5619 - FRIEND, D.J.C., DEPUTY, J., QUEDADO, R.:Photosynthetic and photomorphogenetic
          effects of high photon flux densities on the flowering of two-long-day plants,
          Anagallis arvensis and *Brassica campestris*. - In: MARCELLE, R., CLIJSTERS, H.,
          Van POUCKE, M. (ed.): Photosynthesis and Plant Development. Pp. 59-72. Dr. W.
          Junk bv Publishers, The Hague - Boston - London 1979.

 5620 - FRITSCHEN, L.J., GAY, L.W.: Environmental Instrumentation. - Springer-Verlag,
          New York - Heidelberg - Berlin 1979.

*5621 - FUJIMOTO, K.: [On growth of regeneration trees and environmental factors in
          selection forests (III) Hydrophysiological conditions and chloroplyll content
          in leaves of Sugi seedlings in the model of group-selection stand.] - Bull.
          Ehime Univ. Forest. 14: 25-34, 1977. [In Jap, ab: E.]

*5622 - GAFF, D.F., LATZ, P.K.: The occurrence of resurrection plants in the Australian
          flora. - Aust. J. Bot. 26: 485-492, 1978.

 5623 - GAFF, D.F., McGREGOR, G.R.: The effect of dehydration and rehydration on the
          nitrogen content of various fractions from resurrection plants. - Biol. Plant.
          21: 92-99, 1979.

 5624 - GALLAGHER, J.N., BISCOE, P.V.: Field studies of cereal leaf growth. III. Barley
          leaf extension in relation to temperature, irradiance, and water potential. -
          J. exp. Bot. 30: 645-655, 1979.

5625 - GALLAGHER, J.N., BISCOE, P.V., WALLACE, J.S.: Field studies of cereal leaf growth. IV. Winter wheat leaf extension in relation to temperature and leaf water status. - J. exp. Bot. 30: 657-668, 1979.

5626 - GAMZIKOVA, O.I., GUDINOVA, L.G.: Rannyaya diagnostika zasukho- i zharoustoĭchivosti pshenitsy. [Early diagnosis of drought resistance and heat tolerance for wheat.] - Fiziol. Biokhim. kul't. Rast. 11: 24-28, 1979. [In R, ab: E.]

5627 - GARANOVICH, I.M., VAS'KO, L.P.: Intensiŭnasts' transpiratsyi i sysuchaya sila ablyapikhi krushynapadobnaĭ va ŭmovakh BSSR. [Transpiration rate and suction force of Hippophaë rhamnoides L. in BSSR.] - Vestsi Akad. Navuk Belarus. SSR. Ser. bival. Navuk 1979(6): 15-19. 1979. [In Belorus. ab: E.]

5628 - GARWOOD, E.A., SINCLAIR, J.: Use of water by six grass species. 2. Root distribution and use of soil water. - J. agr. Sci. 93: 25-35, 1979.

5629 - GARWOOD, E.A., TYSON, K.C., SINCLAIR, J.: Use of water by six grass species. 1. Dry-matter yields and response to irrigation. - J. agr. Sci. 93: 13-24, 1979.

*5630 - GAUHL, E.: Photosynthesis of intact leaves and isolated chloroplasts of ecotypes adapted to contrasting light climates. - In: COOMBS, J. (ed.): 4th International Congress on Photosynthesis. P. 127. UKISES, London 1977.

5631 - GAUHL, E.: Sun and shade ecotypes of Solanum dulcamara L.: Photosynthetic light dependence characteristics in relation to mild water stress. - Oecologia 39: 61-70, 1979.

*5632 - GENEROZOVA, I.P., KRIVOVA, I.A.: Osobennosti vosstanovleniya ul'trastruktury khloroplastov posle obezvozhivaniya i posleduyushchego ovodneniya rasteniĭ. [Specialities in regeneration of chloroplats ultrastructure after dehydration and rehydration of plants.] - In: Elektronnaya Mikroskopiya v Botanicheskikh Issledovaniyakh. Pp. 97-99. Zinatne, Riga 1978. [In R.]

5633 - GENKEL', P.A.: Zasukhoustoĭchivost' i produktivnost' rasteniĭ. [Drought resistance and productivity of plants.] - Sel'skokhoz. Biol. 14: 316-322, 1979. [In R, ab: E.]

5634 - GENKEL', P.A., SATAROVA, N.A., SHAPOSHNIKOVA, S.V.: Funktsional'naya aktivnost' khromatina u zakalennykh k zasukhe prorostkov pshenitsy. [Functional activity of chromatin in drought-hardened wheat seedlings.] - Fiziol. Rast. 26: 422-427, 1979. [In R, ab: E.]

*5635 - GENKEL', P.A., TVORUS, E.K.: Razlichiya v sedimentatsii ribosom iz zarodysheĭ nezakalennykh i zakalennykh semyan pshenitsy. [Differences in sedimentation of ribosomes from the embryos of non-hardened and hardened wheat seeds.] - Fiziol. Rast. 25: 236-242, 1978. [In R, ab: E.]

*5636 - GENTRY, H.S., SAUCK, J.R.: The stomatal complex in Agave: Groups Deserticolae, Campaniflorae, Umbelliflorae. - Proc. California Acad. Sci., Ser. 4 41: 371-387, 1978.

5637 - GERAKIS, P.A., PAPAKOSTA-TASOPOULOU, D.: Growth dynamics of Zea mays L. populations differing in genotype and density and grown under illuminance stress. - Oecol. Plant. 14: 13-26, 1979.

5638 - GERBAUD, A., ANDRE, M.: Photosynthesis and photorepiration in whole plants of wheat. - Plant Physiol. 64: 735-738, 1979.

5639 - GERDENITSCH, W.: Mikroskopische Beiträge zum Druck-Volumen-Schema des Zellwasserhaushaltes anhand von Einzelzellen und Zellfäden. - Protoplasma 99: 79-97, 1979.

5640 - GIFFORD, R.M.: Growth and yield of CO_2-enriched wheat under water-limited conditions. - Aust. J. Plant Physiol. 6: 367-378, 1979.

5641 - GIFFORD, R.M.: Carbon dioxide and plant growth under water and light stress: Implications for balancing the global carbon budget. - Search 10: 316-318, 1979.

5642 - GIGON, A.: CO_2-gas exchange, water relations and convergence of mediterranean shrub-types from California and Chile. - Oecol. Plant. 14: 129-150, 1979.

5643 - GILBERT, R.G., NAKAYAMA, F.S., BUCKS, D.A.: Trickle irrigation: Prevention of clogging. - Trans. ASAE 22: 514-519, 1979.

*5644 - GILISSEN, L.J.W.: Post-X-irradiation effects on *Petunia* pollen germinating *in vitro* and *in vivo*. - Environ. exp. Bot. 18: 81-86, 1978.

5645 - GILL, K.S.: Effect of soil salinity on grain filling and grain development in barley. - Biol. Plant. 21: 241-244, 1979.

*5646 - GIMMLER, H., KÜHNL, E.M., CARL, G.: Salinity dependent resistance of *Dunaliella parva* against extreme temperatures. I. Salinity and thermoresistance. - Z. Pflanzenphysiol. 90: 133-153, 1978.

*5647 - GINZBURG, B.Z.: Regulation of cell volume and osmotic pressure in *Dunaliella*. - In: CAPLAN, S.R., GINZBURG, M. (ed.): Energetics and Structure of Halophilic Microorganisms. Pp. 543-560. Elsevier / North-Holland Biomedical Press, Amsterdam - New York 1978.

5648 - GIORDANO, P.M.: Soil temperature and nitrogen effects on response of flooded and nonflooded rice to zinc. - Plant Soil 52: 365-372, 1979.

5649 - GIURGEVICH, J.R., DUNN, E.L.: Seasonal patterns of CO_2 and water vapor exchange of the tall and short height forms of *Spartina alterniflora* Loisel in a Georgia salt marsh. - Oecologia 43: 139-156, 1979.

*5650 - GOAS, M., GOAS, G., LARHER, F.: Quelques aspects du métabolisme des imino-acides chez les halophytes. - Bull. Soc. bot. France 125 (Actualités bot. 3/4): 259-268, 1978.

*5651 - GODULYAN, I.S., SOKRUTA, I.F.: Vlagoobespechennost' i vodopotreblenie kukuruzy posle raznykh predshestvennikov v yugo-vostochnykh raĭonakh stepi Ukrainy. [Water supply and water use by maize after different previous crops in the south-east Ukraine.] - Byull. vses. nauch.-issled. Inst. Kukuruzy 1975(2): 25-28, 1975. [In R.]

5652 - GOL'DFEL'D, M.G., VOZVYSHAEVA, L.V., YUSHMANOV, V.E.: Magnitnaya relaksatsiya protonov vody i sostoyanie sistemy fotorazlozheniya vody v khloroplastakh. [Magnetic relaxation of water protons and the state of water photodissociation system in chloroplasts.] - Biofizika 24: 264-269, 1979. [In R, ab: E.]

*5653 - GOMM, F.B.: Growth and development of meadow plants as affected by environmental variables. - Agron. J. 70: 1061-1065, 1978.

5654 - GONCHAROVA, È.A., UDOVENKO, G.V., YAKOVLEV, A.F.: Osobennosti transporta veshchestv u rasteniĭ pri raznoĭ vodoobespechennosti. [Peculiarities of substances transport in plants with different degree of water supply.] - Fiziol. Biokhim. kul't. Rast. 11: 358-364, 1979. [In R, ab: E.]

*5655 - GORDON, L.Kh.: Dykhanie i Vodno-Solevoĭ Obmen Rastitel'nykh Tkaneĭ. [Respiration and Exchange of Water and Minerals in Plant Tissues.] - Nauka, Moskva 1976. [In R.]

*5656 - GÖRING, H., BLEISS, W., KRETSCHMER, H.: Stimulated elongation growth of coleoptile segments as a consequence of activated H^+ secretion after temporary turgor reduction. - Biochem. Physiol. Pflanz. 173: 373-376, 1978.

5657 - GÖRING, H., POLEVOY, V.V., STAHLBERG, R., STUMPE, G.: Depolarization of trans-
       membrane potential of corn and wheat coleoptiles under reduced water potential
       and after IAA application. - Plant Cell Physiol. 20: 649-656, 1979.

5658 - GÖRING, H., ZOGLAUER, K.: Dependence of IAA-induced ethylene production on
       water potential in excised coleoptile sections of *Triticum aestivum* L. -
       Biochem. Physiol. Pflanz. 174: 568-578, 1979.

5659 - GOUDRIAAN, J., AJTAY, G.L.: The possible effects of increased CO_2 on photo-
       synthesis. - In: BOLIN, B., DEGENS, E.T., KEMPE, S., KETNER, P. (ed.): The
       Global Carbon Cycle. Pp. 237-249. John Wiley & Sons, Chichester 1979.

*5660 - GRADCHANINOVA, O.D.: Metodika anatomicheskogo issledovaniya lista pshenitsy
        v svyazi s fotosintezom. [Methods for anatomical studies of a wheat leaf
        associated with its photosynthetic function.]- Tr. priklad. Bot. Genet. Selek.
        61(3): 68-71, 1978. [In R, ab: E.]

*5661 - GRANDIN, M.: Action complexe du sel et du froid sur la croissance, la multi-
        plication végétative et la floraison de *Glaux maritima* L. - Bull. Soc. bot.
        France 125 (Actualités bot. 3/4): 37-43, 1978.

5662 - GREEN, W.N., FERRIER, J.M., DAINTY, J.: Direct measurement of water capacity
       of *Beta vulgaris* storage tissue sections using a displacement transducer and
       resulting values for cell membrane hydraulic conductivity. - Can. J. Bot. 57:
       981-985, 1979.

5663 - GREENWAY, H., SETTER, T.L.: Na^+, Cl^- and K^+ concentrations in *Chlorella emer-
       sonii* exposed to 100, and 335 mM NaCl. - Aust. J. Plant Physiol. 6: 61-67,
       1979.

5664 - GREENWAY, H., SETTER, T.L.: Accumulation of proline and sucrose during the
       first hours after transfer of *Chlorella emersonii* to high NaCl. - Aust. J.
       Plant Physiol. 6: 69-79, 1979.

5665 - GREGORY, P.J.: Uptake of N, P and K by irrigated and unirrigated pearl millet
       (*Pennisetum typhoides*). - Exp. Agr. 15: 217-223, 1979.

5666 - GREGORY, P.J., SQUIRE, G.R.: Irrigation effects on roots and shoots of pearl
       millet (*Pennisetum typhoides*). - Exp. Agr. 15: 161-168, 1979.

5667 - GRETZMACHER, R.: Die Beeinflussung des morphologischen Ertragsaufbaues und
       der Ertragsleistung durch den Standraum bei Körnermais (*Zea mays* L.). -
       Bodenkultur 30: 256-280, 1979.

5668 - GRIFFIN, D.M., LUARD, E.J.: Water stress and microbial ecology. - In: SHILO, M.
       (ed.): Strategies of Microbial Life in Extreme Environments. Pp. 49-63. Verlag
       Chemie, Weinheim - New York 1979.

5669 - GRIMME, H., NÉMETH, K.: The evaluation of soil K status by means of soil test-
       ing. - In: Potassium Research - Review and Trends. Pp. 99-108. International
       Potash Research Institute, Bern 1979.

5670 - GROSS, J., LENZ, F.: The effects of an induced water stress on pigment changes
       in broccoli (*Brassica oleracea* var *italica*). - Gartenbauwissenschaft 44: 159
       -161, 1979.

5671 - GROSS, P.: Etude et mise au point d'un évaporomètre Piche enregistreur. - Ann.
       Sci. forest. 36: 341-345, 1979.

5672 - GUBBELS, G.H.: Yield and weight per seed in buckwheat after foliar applica-
       tions of growth regulators and antitranspirants. - Can. J. Plant Sci. 59:
       857-859, 1979.

*5673 - GUDERIAN, R.: Air Pollution. (Ecological studies Vol. 22.). - Springer-Verlag,
        Berlin - Heidelberg - New York 1977.

5674 - GULYAEV, B.I.: Reaktsiya ust'its na izmenenie intensivnosti sveta i kontsen-
tratsii CO_2.[Stomata response to changes in light intensity and CO_2 concen-
tration]. - Fiziol. Biokhim. kul't. Rast. 11: 593-600, 1979. [In R, ab: E.]

5675 - GULYÁS, A., BARNA, B., KLEMENT, Z., FARKAS, G.L.: Effect of plasmolytica on the
hypersensitive reaction induced by bacteria in tobacco: a comparison with the
virus-induced hypersensitive reaction. - Phytopathology 69: 121-124, 1979.

5676 - GUPTA, P., SHEORAN, I.S.: Effect of water stress on the enzymes of nitrate
metabolism in two *Brassica* species. - Phytochemistry 18: 1881-1882, 1979.

5677 - GUR, A., DASBERG, S., SCHKOLNIK, I., SAPIR, E., PELED, M.: The influence of
method and frequency of irrigation on soil aeration and some biochemical
responses of apple trees. - Irrig. Sci. 1: 125-134, 1979.

5678 - GUROVICH, L.A.: Effects of improved field practice on crop yield, water use and
profitability of irrigation in central Chile. - Irrig. Sci. 1: 97-105, 1979.

5679 - GUSTA, L.V., FOWLER, D.B., CHEN, P., RUSSELL, D.B., STOUT, D.G.: A nuclear mag-
netic resonance study of water in cold-acclimating cereals. - Plant Physiol.
63: 627-634, 1979.

5680 - GUTTENBERGER, H., HÄRTEL, O., THALER, I.: Scheiden chronisch SO_2-geschädigte
Fichtennadeln Äthylen aus? - Phyton 19: 269-279, 1979.

5681 - HAAS, K., SCHÖNHERR, J.: Composition of soluble cuticular lipids and water
permeability of cuticular membranes from *Citrus* leaves. - Planta 146: 399-403,
1979.

*5682 - HABERMANN, H.M., SHOEMAKER, E.W.: Water utilization, metabolism and stomatal
function in copper deficient *Helianthus annuus*. - Plant Physiol. 61(Suppl.):
88, 1978.

5683 - HABOVŠTIAK, J., TOMKA, O.: Vplyv zrážkových pomerov počas 16-ročného hnojenia
trávnych porastov na ich úrody. [Effect of rainfall on the yields of grassland
fertilized for 16 years.] - Rost. Výroba (Praha) 25: 265-270, 1979. [In Slov,
ab: R,E,G.]

5684 - HAGEMAN, R.H.: Integration of nitrogen assimilation in relation to yield. -
In: HEWITT, E.J., CUTTING, C.V. (ed.): Nitrogen Assimilation of Plants. Pp.
591-611. Academic Press, London - New York - San Francisco 1979.

*5685 - HAGEMANN, R.W., EHLIG, C.F., HUBER, M.J., REYNOSO, R.Y., WILLARDSON, L.S.:
Effect of irrigation frequencies on alfalfa seed yield. - California Agr. 32
(10): 17-18, 1978.

5686 - HALL, A.E., FOSTER, K.W., WAINES, J.G.: Crop adaptation to semi-arid environ-
ments. - In: HALL, A.E., CANNELL, G.H., LAWTON, H.W. (ed.): Agriculture in
Semi-Arid Environments. Pp. 148-179. Springer-Verlag, Berlin - Heidelberg -
New York 1979.

*5687 - HALLAM, N.D., GAFF, D.F.: Re-organization of fine structure during rehydration
of desiccated leaves of *Xerophyta villosa*. - New Phytol. 81: 349-355, 1978.

*5688 - HALLAM, N.D., GAFF, D.F.: Regeneration of chloroplasts structure in *Talbotia
elegans*: a desiccation-tolerant plant. - New Phytol. 81: 657-662, 1978.

*5689 - HALLMAN, E., HARI, P., RÄSÄNEN, P.K., SMOLANDER, H.: Effect of planting shock
on the transpiration, photosynthesis, and height increment of Scots pine seedlings. - Acta forest. fenn. 161: 4-26, 1978.

5690 - HAMAMURA, K.: Varietal differences in ethylene production and etiolated seedling characters in rice with special reference to floating habit. - Jap. J.
Crop Sci. 48: 187-194, 1979.

5691 - HAMAMURA, K.: Comparison of the second leaf blade length between floating and non-floating rice varieties and lines in Central Thailand. - Jap. J. Crop Sci. 48: 201-205, 1979.

5692 - HAMAMURA, K., KUPKANCHANAKUL, T.: Inheritance of floating ability in rice. - Jap. J. Breed. 29: 211-216, 1979.

*5693 - HAMZA, M.: Influence des conditions climatiques et du régime d'apport du NaCl au milieu sur les limites de tolérance d'une espèce résistante: l'*Hedysarum carnosum* Desf. - Bull. Soc. bot. France 125 (Actualités bot. 3/4): 45-51, 1978.

*5694 - HAMZA, M.: Influence du régime d'apport du NaCl au milieu sur la régulation du bilan hydrique et de la teneur ionique chez une espèce tolérante, l'*Hedysarum carnosum* Desf., et une espèce sensible, le Haricot, *Phaseolus vulgaris* L. - Bull. Soc. bot. France 125 (Actualités bot. 3/4): 177-187, 1978.

5695 - HANCOCK, J.G.: Association of an irreversible reduction in urea permeability with osmotic factors during the ageing of excised squash hypocotyls. - Can. J. Bot. 57: 26-33, 1979.

5696 - HANSEN, E.A., DICKSON, R.E.: Water and mineral nutrient transfer between root systems of juvenile *Populus*. - Forest Sci. 25: 247-252, 1979.

5697 - HANSEN, V., HEGG, K.: Description of a computer-controlled data logging system for measuring meteorological variables. - Agr. Meteorol. 20: 451-457, 1979.

5698 - HANSON, A.D., TULLY, R.E.: Amino acids translocated from turgid and water-stressed barley leaves II. Studies with ^{13}N and ^{14}C. - Plant Physiol. 64: 467-471, 1979.

5699 - HANSON, A.D., TULLY, R.E.: Light stimulation of proline synthesis in water-stressed barley leaves. - Planta 145: 45-51, 1979.

5700 - HANSON, W.D., YEH, R.Y.: Genotypic differences for reduction in carbon exchange rates as associated with assimilate accumulation in soybean leaves. - Crop Sci. 19: 54-58, 1979.

*5701 - HARDT, S.: Aspects of diffusional transport in microorganisms. - In: CAPLAN, S.R., GINZBURG, M. (ed.): Energetics and Structure of Halophilic Microorganisms. Pp. 591-597. Elsevier / North-Holland Biomedical Press, Amsterdam - New York 1978.

5702 - HARDY, R.W.F.: Chemical plant growth regulation in world agriculture. - In: SCOTT, T.K. (ed.): Plant Regulation and World Agriculture. Pp. 165-206. Plenum Press, New York - London 1979.

5703 - HARI, P., KANNINEN, M., KELLOMÄKI, S., LUUKKANEN, O., PELKONEN, P., SALMINEN, R., SMOLANDER, H.: An automatic system for measurements of gas exchange and environmental factors in a forest stand, with special reference to measuring principles. - Silva fenn. 13: 94-100, 1979.

5704 - HARLEY, J.L., SCOTT RUSSELL, R. (ed.): The Soil-Root Interface. (Proceedings of an International Symposium Held in Oxford, England, March 28 to 31, 1978). - Academic Press, London - New York - San Francisco 1979.

5705 - HARNISCHFEGER, G.: Connection between the rate of cooling and fluorescence properties at 77 K of isolated chloroplasts. - Biochim. biophys. Acta 546: 348-355, 1979.

5706 - HART, F.X., SCHOTTENFELD, R.S.: Evaporation and plant damage in electric fields. - Int. J. Biometeorol. 23: 63-68, 1979.

5707 - HASSANYAR, A.S., WILSON, A.M.: Tolerance of desiccation in germinating seeds of crested wheatgrass and Russian wildrye. - Agron. J. 71: 783-786, 1979.

5708 - HATFIELD, J.L., CARLSON, R.E.: Leaf-water-potential patterns within two soybean cultivars. - Iowa State J. Res. 53: 259-267, 1979.

5709 - HAUPT, W., FEINLEIB, M.E. (ed.): Physiology of Movements.(Encyclopedia of Plant Physiology, New Series, Volume 7). - Springer-Verlag, Berlin - Heidelberg - New York 1979.

*5710 - HAUTECLOQUE, J.: Exploitation des données psychrométriques à l'aide d'une calculatrice - expérience d'un observateur. - Bull. inform. Agrometeorol. 1978 (2): 32-34, 1978.

*5711 - HEAL, O.W., PERKINS, D.F. (ed.): Production Ecology of British Moors and Montane Grasslands. - Springer-Verlag, Berlin - Heidelberg - New York 1978.

5712 - HEATHCOTE, D.G., ETHERINGTON, J.R., WOODWARD, F.I.: Instrument for non-destructive measurement of the pressure potential (turgor) of leaf cells. - J. exp. Bot. 30: 811-816, 1979.

5713 - HEATHERLY, L.G., RUSSELL, W.J.: Effect of soil water potential of two soils on soybean emergence. - Agron. J. 71: 980-982, 1979.

5714 - HEBBLETHWAITE, P.D., McLAREN, J.S.: Nitrogen studies in *Lolium perenne* grown from seed. III. The effect of nitrogen and water stress. - Grass Forage Sci. 34: 221-227, 1979.

*5715 - HEEMST, H.D.J., van, KEULEN, H., van, STOLWIJK, H.: Potentiële Produktie, Bruto- en Nettoproduktie van de Nederlandse Landbow. [Potential, Gross and Net Production of Netherlands Agriculture.] - Pudoc, Wageningen 1978. [In Dutch, ab: E.]

5716 - HEGARTY, T.W., ROSS, H.A.: Effects of light and growth regulators on germination and radicle growth of lettuce seeds held under high-temperature stress and water stress. - New Phytol. 82: 49-57, 1979.

5717 - HEGARTY, T.W., ROSS, H.A.: Use of growth regulators to remove the differential sensitivity to moisture stress of seed germination and seedling growth in red clover (*Trifolium pratense* L.). - Ann. Bot. 43: 657-660, 1979.

*5718 - HEGDE, B.R., HAVANAGI, G.V., MUNIKRISHNA REDDY, N.: Fertilizer need and water use efficiency of dryland ragi (*Eleusine coracana*) in a cowpea-ragi sequence. - Mysore J. agr. Sci. 12: 433-436, 1978.

*5719 - HEGDE, B.R., HAVANAGI, G.V., SATYANARAYANA, T., NANJUNDA REDDY, S., BADANUR, V.P.: Consumptive use of water by dryland crops in Bangalore region. - Ind. J. Agr. 23: 87-92, 1978.

*5720 - HEIM, G., GROUZIS, M.: Halophilie et résistance au sel chez deux Salicornes annuelles du littoral méditerranéen. - Bull. Soc. bot. France 125 (Actualités bot. 3/4): 61-69, 1978.

5721 - HEINZE, M., FIEDLER, H.J.: Versuche zur Begrünung von Kalirückstandshalden. 1. Mitteilung: Gefässversuche mit Bäumen und Sträuchern bei unterschiedlichen Wasser- und Nährstoffangebot. - Arch. Acker- Pflanzenbau Bodenk. 23: 315-322, 1979.

*5722 - HEKMAT-SHOAR, H.: Analyse des relations phylogéniques entre trois *Suaeda* par culture en milieux diversement salés. - Bull. Soc. bot. France 125 (Actualités bot. 3/4): 293-306, 1978.

5723 - HELLALI, R., KESTER, D.E.: High temperature induced bud-failure symptoms in vegetative buds of almond plants in growth chambers. - J. Amer. Soc. hort. Sci. 104: 375-378, 1979.

5724 - HELLALI, R., KESTER, D.E., MARTIN, G.C.: Seasonal changes in the physiology of vegetative buds of normal and bud-failure affected 'Nonpareil' almond trees. - J. Amer. Soc. hort. Sci. 104: 371-375, 1979.

*5725 - HERLIHY, M.: Dry matter response of ryegrass to ammonium and nitrate sources of nitrogen as a function of soil texture and moisture. - Plant Soil 50: 633-646, 1978.

*5726 - HERLIHY, M., McALEESE, D.M.: Nitrogen uptake efficiency by ryegrass in soils of a textural sequence 1. Effect of soil moisture availability. - Irish J. agr. Res. 17: 61-70, 1978.

5727 - HINCKLEY, T.M., DOUGHERTY, P.M., LASSOIE, J.P., ROBERTS, J.E., TESKEY, R.O.: A severe drought: Impact on tree growth, phenology, net photosynthetic rate and water relations. - Amer. Midl. Natur. 102: 307-316, 1979.

5728 - HIRASAWA, T., ISHIHARA, K.: [The relationship between environmental factors and water status in the rice plant. II. On leaf water potential and xylem water potential.] - Jap. J. Crop Sci. 48: 557-568, 1979. [In Jap, ab: E.]

5729 - HO, L.C., SHAW, A.F.: Net accumulation of minerals and water and the carbon budget in an expanding leaf of tomato. - Ann. Bot. 43: 45-54, 1979.

5730 - HODDINOTT, J., EHRET, D.L., GORHAM, P.R.: Rapid influences of water stress on photosynthesis and translocation in *Phaseolus vulgaris*. - Can. J. Bot. 57: 768-776, 1979.

*5731 - HOLDSWORTH, E.S., BRUCK, K.: Enzymes concerned with β-carboxylation in marine phytoplankter. Purification and properties of phosphoenolpyruvate carbokinase. - Arch. Biochem. Biophys. 182: 87-94, 1977.

5732 - HOOPER, A.W., SMITH, R.A., BOWMAN, G.E.: A near-infrared diffuse reflectance spectrophotometer. - J. agr. eng. Res. 24: 79-85, 1979.

*5733 - HORIE, T.: Studies on photosynthesis and primary production of rice plants in relation to meteorological environments. 1. Gaseous diffusive resistances, photosynthesis and transpiration in the leaves as influenced by radiation intensity and wind speed. - J. agr. Meteorol. 34: 125-136, 1978.

5734 - HORST, G.L., NELSON, C.J.: Compensatory growth of tall fescue following drought - Agron. J. 71: 559-563, 1979.

5735 - HORTON, R.F.: Hormonal regulation of stomatal function. - In: SEN, D.N., CHAWAN, D.D., BANSAL, R.P. (ed.): Structure, Function and Ecology of Stomata. Pp. 103-120. Bishen Singh Mahendra Pal Singh, Dehra Dun 1979.

5736 - HUANG, A.H.C., CAVALIERI, A.J.: Proline oxidase and water stress-induced proline accumulation in spinach leaves. - Plant Physiol. 63: 531-535, 1979.

5737 - HUBER, W., SANKHLA, N.: Effect of sodium chloride on photosynthesis of *Lemna minor* L. - Z. Pflanzenphysiol. 91: 147-156, 1979.

5738 - HUBER, W., WÖRLE, I., SANKHLA, N.: Effects of kinetin and proline on enzyme activities in roots of waterlogged *Pisum sativum* plants. - Flora 168: 217-226, 1979.

5739 - HUCK, M.G.: A photographic view of microscopic processes at the root-soil interface. - In: HARLEY, J.L., SCOTT RUSSELL, R. (ed.): The Soil-Root Interface. Pp. 273-274. Academic Press, London - New York - San Francisco 1979.

5740 - HUISINGA, B.: Control of loading and unloading by turgor regulation in long distance transport. - Acta Bot. Neerl. 28: 67-72, 1979.

*5741 - HUKKERI, S.B., SHARMA, A.K., NIMBOLE, N.N., BASANTANI, H.T.: Stress-day index
for timing of irrigation for potato. - Ind. J. agr. Sci. 45: 515-523, 1975.

*5742 - HULL, H.M., MORTON, H.L., WHARRIE, J.R.: Environmental influences on cuticle
development and resultant foliar penetration. - Bot. Rev. 41: 421-452, 1975.

5743 - HULL, H.M., WENT, F.W., BLECKMANN, C.A.: Environmental modifications of epicuticular wax structure of Prosopis leaves. - J. Arizona - Nevada Acad. Sci.
14: 39-42, 1979.

5744 - HUMMEL, R.L., PELLETT, H., PARSONS, L.: Variability associated with leaf water
potential in red-osier dogwood in field and growth chamber environments. -
Can. J. Plant Sci. 59: 847-852, 1979.

5745 - HUZULÁK, J.: The xylem pressure potential of shrub species in an oak-hornbeam
forest. - Biol. Plant. 21: 9-14, 1979.

*5746 - HUZULÁK, J., ELIÁŠ, P.: Príspevok k štúdiu vodného režimu Carpinus betulus.
[A contribution to the study of water regime of Carpinus betulus.] - Acta Bot.
Slov. Acad. Sci. Slov., Ser. A 4: 277-286, 1978. [In Slov, ab: R,E.]

5747 - IBRAHIM, M., BERGER, A., RAPP, M.: Determination de la transpiration au moyen
de l'eau tritiée. Méthodologie et validité des résultats. Application a Pinus
pinea L. - Plant Soil 52: 291-301, 1979.

*5748 - IMAMALIEV, A.I., NASYROVA, T.Kh.: Vliyanie gerbitsidov na nekotorye fiziologo-
-biokhimicheskie protsessy khlopchatnika v zavisimosti ot vlazhnosti pochvy.
[Effect of herbicides on some physiological and biochemical processes of
cotton depending on soil moisture.] - Dokl. VASCHNIL 1975(5): 19-21, 1975.
[In R.]

*5749 - INGOLD, C.T.: Water and spore liberation. - In: KOZLOWSKI, T.T. (ed.): Water
Deficits and Plant Growth. Volume V. Pp. 119-140. Academic Press, New York -
San Francisco - London 1978.

*5750 - ISAAKIDOU, J., PAPAGEORGIOU, G.: Effects of imidoester crosslinking on structu-
ral and functional characteristics of isolated spinach chloroplasts. - Biophys.
J. 16(No. 2, Part 2): 161, 1976.

*5751 - ISAAKIDOU, J., PAPAGEORGIOU, G.: Molecular and functional properties of isola-
ted chloroplasts after crosslinking with dimethylsuberimidate. - In: COOMBS,
J. (ed.): 4th International Congress on Photosynthesis. P. 175. UKISES, London
1977.

5752 - ISFAN, D.: Nitrogen rate-yield-precipitation relationships and N rate fore-
casting for corn crops. - Agron. J. 71: 1045-1051, 1979.

5753 - ISHIHARA, K.: Diurnal course of stomatal aperture of leaf blades in rice
plants. - JARQ 13: 85-89, 1979.

5754 - ISHIHARA, K., HIRASAWA, T., IIDA, O., OGURA, T.: [An improved infiltration
method for measuring the narrow stomatal aperture of leaf blades in rice
plants.] - Jap. J. Crop Sci. 48: 319-320, 1979. [In Jap.]

5755 - ISHIHARA, K., IIDA, O., HIRASAWA, T., OGURA, T.:[Relationship between nitrogen
content in leaf blades and photosynthetic rate of rice plants with reference
to stomatal aperture and conductance.]- Jap. J. Crop Sci. 48: 543-550, 1979.
[In Jap, ab: E.]

5756 - ISHIHARA, K., KURODA, E., ISHII, R., OGURA, T.: [Relationship between nitrogen
content in leaf blades and photosynthetic rate in rice plants measured with
an infrared gas analyzer and an oxygen electrode.] - Jap. J. Crop Sci. 48:
551-556, 1979. [In Jap, ab: E.]

*5757 - ITAI, C.: Response of *Eucalyptus occidentalis* to water stress induced by NaCl.
        - Physiol. Plant. 43: 377-379, 1978.

 5758 - IVANCHENKO, V.M.: Regulyatsiya funktsiĭ membran sformirovannykh khloroplastov
        i intensivnost' fotosinteza. [Regulation of functions of membranes of developed
        chloroplasts and photosynthetic rate.] - In: GONCHARIK, M.N. (ed.): Regulyatsiya
        Funktsiĭ Membran Rastitel'nykh Kletok. Pp. 147-197. Nauka i Tekhnika, Minsk
        1979. [In R.]

*5759 - IVANOVA, N.A.: Vliyanie defoliatsii na stroenie ust'ichnogo apparata i foto-
        sinteticheskuyu aktivnost' list'ev. [Influence of defoliation on the structu-
        re of stomatic apparatus and leaf photosynthetic activity.] - In: Mezostruktu-
        ra i Funktsional'naya Aktivnost' Fotosinteticheskogo Apparata. Pp. 132-136.
        Ural'skiĭ Gosudarstvennyĭ Universitet, Sverdlovsk 1978. [In R.]

 5760 - IWAI, S., KAWASHIMA, N., MATSUYAMA, S.: Effect of water stress on proline
        catabolism in tobacco leaves. - Phytochemistry 18: 1155-1157, 1979.

 5761 - JACKSON, M.B.: Rapid injury to peas by soil waterlogging. - J. Sci. Food Agr.
        30: 143-152, 1979.

 5762 - JACKSON, P.A., SPOMER, G.G.: Biophysical adaptations of four western conifers
        to habitat water conditions. - Bot. Gaz. 140: 428-432, 1979.

 5763 - JAFFEE, B.A., MAI, W.F.: Effect of soil water potential on growth of apple
        trees infected with *Pratylenchus penetrans*. - J. Nematol. 11: 165-168, 1979.

*5764 - JAIN, V.K., MUKHERJEE, D.: Effect of morphactin on leaf epidermis of *Lycoper-
        sicon esculentum* Mill. - Curr. Sci. 47: 811-812, 1978.

 5765 - JANERETTE, C.Λ.: The effects of water soaking on the germination of sugar
        maple seeds. - Seed Sci. Technol. 7: 341-346, 1979.

 5766 - JANERETTE, C.A.: The pathway of water entry into sugar maple seeds. - Seed
        Sci. Technol. 7: 347-353, 1979.

 5767 - JANERETTE, C.A.: Seed dormancy in sugar maple. - Forest Sci. 25: 307-311, 1979.

*5768 - JARVIS, P.G., JAMES, G.B., LANDSBERG, J.J.: Coniferous forest. - In: MONTEITH,
        J.L. (ed.): Vegetation and the Atmosphere. Volume 2. Case Studies. Pp. 171-240.
        Academic Press, London - New York - San Francisco 1976.

*5769 - JEANRENAUD, E.: La dynamique de l'intensité de la respiration et ses relations
        avec l'intensité de la transpiration et avec le déficit hydrique, chez quelques
        espèces de plantes. - Rev. Roum. Biol., Sér. Biol. vég. 23: 145-158, 1978.

 5770 - JEFFERIES, R.L., RUDMIK, T., DILLON, E.M.: Responses of halophytes to high sa-
        linities and low water potentials. - Plant Physiol. 64: 989-994, 1979.

 5771 - JEJE, A.A., ZIMMERMANN, M.H.: Resistance to water flow in xylem vessels. -
        J. exp. Bot. 30: 817-827, 1979.

 5772 - JENKINS, G., RHODES, A.P., GILL, A.A., HANSON, P.R.: The effect of irrigation
        and nitrogen supply on the yield and quality of protein in high-lysine barleys.
        J. Sci. Food Agr. 30: 647-652, 1979.

*5773 - JENSEN, I., MEHLSEN, J.: The influence of various watering methods on plant
        growth and nutrient uptake in pot experiments. - Kongel. Vet. Landbohøjsk.
        Årsskrift 1978: 1-12, 1978.

 5774 - JERNSTEDT, J.A., CLARK, C.: Stomata on the fruits and seeds of *Eschscholtzia*
        (*Papaveraceae*). - Amer. J. Bot. 66: 586-590, 1979.

5775 - JOHANSSON, M., WÄSTERLUND, I.: Root and transpiration studies on young Norway spruce trees with die back symptoms in Sweden. - Europ. J. Forest Pathol. 9: 257-264, 1979.

5776 - JOHNSON, A.W., SUMNER, D.R., JAWORSKI, C.A.: Effect of film mulch, trickle irrigation, and DD-MENCS on nematodes, fungi, and vegetable yields in a multi-crop production system. - Phytopathology 69: 1172-1175, 1979.

5777 - JOHNSON, C.R., NELL, T.A., JOINER, J.N., KRANTZ, J.K.: Effects of light intensity and potassium on leaf stomatal activity of *Ficus benjamina* L. - Hort-Science 14: 277-278, 1979.

*5778 - JOHNSON, D.A., ASAY, K.H.: A technique for assessing seedling emergence under drought stress. - Crop Sci. 18: 520-522, 1978.

5779 - JOHNSON, D.W., KUNTZ, J.E.: *Eutypella* canker of maple: Ascospore discharge and dissemination. - Phytopathology 69: 130-135, 1979.

5780 - JOHNSON, H.B., ROWLANDS, P.G., TING, I.P.: Tritium and carbon-14 double iso-tope porometer for simultaneous measurements of transpiration and photosynthe-sis. - Photosynthetica 13: 409-418, 1979.

5781 - JOHNSSON, A., BROGÅRDH, T., HOLJE, Ø.: Oscillatory transpiration of *Avena* plants: Perturbation experiments provide evidence for a stable point of sin-gularity. - Physiol. Plant. 45: 393-398, 1979.

5782 - JOHNSTON, S.K., WALKER, R.H., MURRAY, D.S.: Germination and emergence of hemp sesbania (*Sesbania exaltata*). - Weed Sci. 27: 290-293, 1979.

5783 - JONES, H.G.: Stomatal behavior and breeding for drought resistance. - In: MUSSELL, H., STAPLES, R. (ed.): Stress Physiology in Crop Plants. Pp. 407-428. John Wiley & Sons, Inc., New York 1979.

5784 - JONES, H.G.: Screening for tolerance of photosynthesis to osmotic and saline stress using rice leaf-slices. - Photosynthetica 13: 1-8, 1979.

5785 - JONES, H.G.: Visual estimation of plant water status in cereals. - J. agr. Sci. 92: 83-89, 1979.

5786 - JONES, H.G., HIGGS, K.H.: Water potential - water content relationships in apple leaves. - J. exp. Bot. 30: 965-970, 1979.

5787 - JONES, H.G., NORTON, T.A.: Internal factors controlling the rate of evapora-tion from fronds of some intertidal algae. - New Phytol. 83: 771-781, 1979.

5788 - JONES, M.M., RAWSON, H.M.: Influence of rate of development of leaf water de-ficit upon photosynthesis, leaf conductance, water use efficiency, and osmotic potential in sorghum. - Physiol. Plant. 45: 103-111, 1979.

5789 - JONES, T.W., GALLOWAY, R.A.: Effect of light quality and intensity on glycerol content in *Dunaliella tertiolecta* (*Chlorophyceae*) and the relationship to cell growth / osmoregulation. - J. Phycol. 15: 101-106, 1979.

*5790 - JORDAN, C.F.: Stem flow and nutrient transfer in a tropical rain forest. - Oikos 31: 257-263, 1978.

5791 - JORDAN, L.S., SHANER, D.L.: Weed control. - In: HALL, A.E., CANNELL, G.H., LAWTON, H.W. (ed.): Agriculture in Semi-Arid Environments. Pp. 266-296. Springer-Verlag, Berlin - Heidelberg - New York 1979.

5792 - JOURNET, A.R.P.: The water status of *Eucalyptus blakelyi* M. under field condi-tions. - Ann. Bot. 44: 125-128, 1979.

5793 - JUNGE, W., AUSLÄNDER, A., McGEER, A.J., RUNGE, T.: The buffering capacity of
       the internal phase of thylakoids and the magnitude of the pH changes inside
       under flashing light. - Biochim. biophys. Acta 546: 121-141, 1979.

5794 - JURY, W.A.: Water transport through soil, plant, and atmosphere. - In: HALL,
       A.E., CANNELL, G.H., LAWTON, H.W. (ed.): Agriculture in Semi-Arid Environ-
       ments. Pp. 180-199. Springer-Verlag, Berlin - Heidelberg - New York 1979.

5795 - KABAKI, N., SAKA, H., AKITA, S.: [Effects of nitrogen, phosphorus and potassium
       deficiencies on photosynthesis and RuBP carboxylase-oxygenase activities in
       rice plants.] - Jap. J. Crop Sci. 48: 378-384, 1979. [In Jap, ab: E.]

5796 - KÁBRT, B.: Ovplyvnenie úrody a troch znakov pšenice ozimnej atmosferickými
       zrážkami a teplotami počas vegetácie. [Effect of temperatures and atmospheric
       precipitations during vegetation on the yield and three parameters of winter
       wheat.] - Rost. Výroba (Praha) 25: 277-287, 1979. [In Slov, ab: R,E,G.]

5797 - KACHEL, K., ROTH, D.: Untersuchungen zum Einfluss der Beregnung auf wichtige
       Qualitätsparameter der Zuckerrübe auf zwei unterschiedlichen Schwarzerdestand-
       orten im Thüringer Becken. - Arch. Acker- Pflanzenbau Bodenk. 23: 199-204,
       1979.

*5798 - KAKHNOVICH, L.V. (ed.): Optimizatsiya Fotosinteticheskogo Apparata Vozdeľstviem
       Razlichnykh Faktorov. [Optimization of the Photosynthetic Apparatus by the
       Action of Different Factors.] - Izd. BGU, Minsk 1976. [In R.]

5799 - KALMA, J.D., FUCHS, M.: Citrus orchards. - In: MONTEITH, J.L. (ed.): Vegetation
       and the Atmosphere. Volume 2. Case Studies. Pp. 309-328. Academic Press, Lon-
       don - New York - San Francisco 1976.

5800 - KAMPRATH, E.J., CASSEL, D.K., GROSS, H.D., DIBB, D.W.: Tillage effects on
       biomass production and moisture utilization by sobeans on coastal plain soils.
       - Agron. J. 71: 1001-1005, 1979.

*5801 - KANEMASU, E.T., RASMUSSEN, V.P., BAGLEY, J.: Estimating Water Requirements for
       Corn with a "Pocket" Calculator. - Agricultural Experiment Station, Kansas
       State University, Manhattan 1978.

*5802 - KANG, M.S., ZUBER, M.S., HORROCKS, R.D.: An electronic probe for estimating
       ear moisture content of maize. - Crop Sci. 18: 1083-1084, 1978.

5803 - KANNABIRAN, B.: Morphology and development of foliar epidermis in *Stylosanthes
       fruticosa* (Retz.) Alston (Syn. *S. mucronata* Willd.). - Proc. Ind. Acad. Sci.
       88 B II (2): 155-160, 1979.

5804 - KANNABIRAN, B., KRISHNAMURTHY, K.H.: Stomatal ontogeny in some dicotyledonous
       families with special reference to its taxonomic and phylogenetic value. -
       In: SEN, D.N., CHAWAN, D.D., BANSAL, R.P. (ed.): Structure, Function and Eco-
       logy of Stomata. Pp. 23-41. Bishen Singh Mahedra Pal Singh, Dehra Dun 1979.

*5805 - KAPAHI, B.K., SARIN, Y.K.: Natural factors governing the growth and alkaloid
       yield in *Datura innoxia* Mill. - Ind. J. Pharmacy 40: 14-15, 1978.

5806 - KAPPEN, L., LANGE, O.L., SCHULZE, E.-D., EVENARI, M., BUSCHBOM, U.: Ecophy-
       siological investigations on lichens of the Negev Desert. VI. Annual course
       of the photosynthetic production of *Ramalina maciformis* (Del.) Bory. - Flora
       168: 85-108, 1979.

5807 - KARAMANOS, A.J.: Water stress: A challenge for the future of agriculture. -
       In: SCOTT, T.K. (ed.): Plant Regulation and World Agriculture. Pp. 415-455.
       Plenum Press, New York - London 1979.

5808 - KARAPETYAN, N.V., BUKHOV, N.G.: Vliyanie degidratatsii na funktsionirovanie
       fotosistem vysshikh rastenii. [The effect of dehydration on functioning of
       photosystems of higher plants.] - Molekul. Biol. 13: 947-954, 1979. [In R, ab: E.]

5809 - KARLIC, H., RICHTER, H.: Storage of detached leaves and twigs without changes in water potential. - New Phytol. 83: 379-384, 1979.

5810 - KARMANOV, V.G., SOLOV'EV, E.V.: Vllyanie pollvov I sveta na intensivnost' dykhaniya kornevykh sistem rastenii. [Effect of irrigation and illumination on respiration rate in plant root systems.] - Fiziol. Biokhim. kul't. Rast. 11: 245-250, 1979. [In R, ab: E.]

*5811 - KARTUSCH, B.: Unterschiedliches Photosyntheseverhalten immergrüner Pflanzen in Abhängigkeit von den klimatischen Faktoren. - Phyton 19: 61-69, 1978.

5812 - KASIMOV, I., DIMITROV, I., PETROV, G., SIMEONOV, B., PETROVA, M.: Vllyanie na produktivnite svoïstva na sortovete, ravnishcheto na torene i napoyavaneto v"rkhu dobiva ot zimnata meka pshenitsa v Dobrudzha. [Variety productive properties, fertilizer level and irrigation as affecting the winter soft wheat yield in Dobrudzha.] - Rasteniev"dni Nauki 16(2): 75-85, 1979. [In Bulg, ab: R,E.]

5813 - KÄSTNER, A.: Beiträge zur Wuchsformanalyse und systematischen Gliederung von *Teucrium* L. II. Anatomie der Sprosse und Blätter. - Flora 168: 431-467, 1979.

5814 - KATAOKA, K., KANEKO, M.:[Varietal difference of stomatal frequency and inter-veinal distance of rice leaf blades.] - Bull. Fac. Agr. Tamagawa Univ. 1979 (19): 79-84, 1979. [In Jap, ab: E.]

5815 - KATERJI, N.B.: Étude comparative de la résistance du couvert végétal au trans-fert de la vapeur d'eau avec le potentiel hydrique des feuilles chez le Blé. - Oecol. Plant. 14: 55-60, 1979.

*5816 - KAUSS, H.: Osmotic regulation in algae. - In: REINHOLD, L., HARBORNE, J.B., SWAIN, T. (ed.): Progress in Phytochemistry. Volume 5. Pp. 1-27. Pergamon Press, Oxford - New York - Toronto - Sydney - Paris - Frankfurt 1978.

5817 - KAUSS, H.: Biochemie der osmotischen Regulation bei *Poterioochromonas malha-mensis*. - Ber. Deut. bot. Ges. 92: 11-22, 1979.

5818 - KAUSS, H., THOMSON, K.S., THOMSON, M., JEBLICK, W.: Osmotic regulation. Phy-siological significance of proteolytic and nonproteolytic activation of iso-floridoside-phosphate synthase. - Plant Physiol. 63: 455-459, 1979.

5819 - KAWAMATA, S.: Studies on a hardened disorder of Japanese pear (*Pyrus serotina* var. *culta* Rehder) fruits. II. Effects of calcium supply and soil moisture on the incidence of 'Yuzuhada' disorder of pot-grown Nijisseiki pear trees. - J. Jap. Soc. hort. Sci. 48: 137-146, 1979.

5820 - KAWASE, M.: Role of cellulase in aerenchyma development in sunflower. - Amer. J. Bot. 66: 183-190, 1979.

*5821 - KAYANI, S.A., SHEIKH, K.H.: Seed germination and growth of *Hibiscus cannabinus* L. at different levels of soil moisture. - Biologia (Lahore) 23: 23-30, 1977.

5822 - KAZAKOV, E.A., OKANENKO, A.S.: Potentsial'nye vozmozhnosti intensivnosti foto-sinteza i produktivnosti sakharnoï svekly v razlichnykh usloviyakh vodnogo rezhima pochvy. [Possibilities of photosynthetic rate and productivity of sugar beet under different conditions of soil water regime.] - Fiziol. Biokhim. kul't. Rast. 11: 574-582, 1979. [In R, ab: E.]

5823 - K"DREV, T., GEORGIEVA, V.: Vllyanie na predseitbenata obrabotka s bor na nya-koi strani ot vodoobmena i nitratredutsirashchata sposobnost na mladi tsare-vichni rasteniya. [Effect of boron pre-seeding treatment on some aspects of water exchange and on the nitrate reducing ability of young maize plants.] - Fiziol. Rast. (Sofia) 5(1): 78-82, 1979. [In Bulg, ab: R,E.]

*5824 - KEELEY, J.E.: Malic acid accumulation in roots in response to flooding: Evidence contrary to its role as an alternative to ethanol. - J. exp. Bot. 29: 1345-1349, 1978.

5825 - KEIM, D.L., KRONSTAD, W.E.: Drought resistance and dryland adaptation in winter wheat. - Crop Sci. 19: 574-576, 1979.

*5826 - KELLOMÄKI, S., HARI, P.: Rate of photosynthesis of some forest mosses as a function of temperature and light intensity and effect of water content of moss cushion on photosynthetic rate. - Silva fenn. 10: 288-295, 1976.

5827 - KENG, J.C.W., SCOTT, T.W., LUGO-LOPEZ, M.A.: Fertilizer management with drip irrigation in an Oxisol. - Agron. J. 71: 971-980, 1979.

5828 - KENNEDY, C.D.: The effect of applied suctions and 2,4-dichlorophenoxyacetic acid on water and salt efflux from the cut ends of excised maize roots. - J. exp. Bot. 30: 275-289, 1979.

*5829 - KENNEDY, R.A.: Variation in C_3 and C_4 characteristics within the genus *Mollugo*. - In: COOMBS, J. (ed.): 4th International Congress on Photosynthesis. P. 192. UKISES, London 1977.

*5830 - KERCHER, J.R.: GROW1: A Crop Growth Model for Assessing Impacts of Gaseous Pollutants from Geothermal Technologies. - University of California, Livermore 1977.

*5831 - KERR, J.P., McPHERSON, H.G.: Evapotranspiration and physiological response to water stress of several pasture and crop species. - Proc. N. Zeal. Grassland Assoc. 39(1): 70-78, 1978.

5832 - KERSHAW, K.A., MORRIS, T., TYSIACZNY, M.J., MacFARLANE, J.D.: Physiological-environmental interactions in lichens. VIII. The environmental control of dark CO_2 fixation in *Parmelia caperata* (L.) Ach. and *Peltigera canina* var. *praetextata* Hue. - New Phytol. 83: 433-444, 1979.

5833 - KESSELL, S.R.: Adaptation and dimorphism in eastern hemlock, *Tsuga canadensis* (L.) Carr. - Amer. Natur. 113: 333-350, 1979.

5834 - KHAN, A.A., KARSEN, C.M., LEUE, E.F., ROE, C.H.: Preconditioning of seeds to improve performance. - In: SCOTT, T.K. (ed.): Plant Regulation and World Agriculture. Pp. 395-413. Plenum Press, New York - London 1979.

5835 - KHFAJI, A.K., NORTON, T.A.: The effects of salinity on the distribution of *Fucus ceranoides*. - Estuarine coastal marine Sci. 8: 433-439, 1979.

5836 - KIMBALL, B.A., MITCHELL, S.T.: Tomato yields from CO_2-enrichment in unventilated and conventionally ventilated greenhouses. - J. Amer. Soc. hort. Sci. 104: 515-520, 1979.

5837 - KIMENOV, G.P., MINKOV, I.N.: Vliyanie vodnogo defitsita na raspredelenie ^{14}C sredi produktov fotosinteza v list'yakh *Haberlea rhodopensis* Friv. I *Ramonda serbica* Panc. [Effect of water deficiency on the ^{14}C-incorporation into the products of photosynthesis in the leaves of *Haberlea rhodopensis* Friv. and *Ramonda serbica* Panc.]- In: VAKLINOVA, S.G., VANKOVA-RADEVA, R., VASILEVA, V.S. (ed.): Fotosinteticheskaya Assimilyatsiya CO_2 I Fotodykhanie. Pp. 92-97. Izdatel'stvo Bolgarskoĭ Akademii Nauk, Sofiya 1979. [In R.]

5838 - KINERSON, R.S.: Studies of photosynthesis and diffusion resistance in paper birch (*Betula papyrifera* Marsh.) with synthesis through computer simulation. - Oecologia 39: 37-49, 1979.

5839 - KING, R.W., SALMINEN, S.O., HILL, R.D., HIGGINS, T.J.V.: Abscisic-acid and gibberellin action in developing kernels of triticale (cv. 6A190). - Planta 146: 249-255, 1979.

5840 - KIRKHAM, M.B.: Effect of FeEDDHA on the water relations of wheat. - J. Plant
       Nutr. 1: 417-424, 1979.

5841 - KIRKHAM, M.B.: Water relations of wheat alternated between two root temperatu-
       res. - In: HARLEY, J.L., SCOTT RUSSELL, R. (ed.): The Soil-Root Interface.
       Pp. 425-426. Academic Press, London - New York - San Francisco 1979.

5842 - KIRKHAM, M.B.: Water relations of wheat alternated between two root temperatu-
       res. - New Phytol. 82: 89-96, 1979.

5843 - KIRKHAM, M.B., GABRIELS, D.: Water and nutrient uptake of wick-grown plants. -
       Hort. Res. 19: 3-13, 1979.

5844 - KIRKHAM, M.B., SMITH, E.L.: Water potential gradients of tall and short culti-
       vars of winter wheat. - Plant Soil 52: 553-559, 1979.

5845 - KIRST, G.O., BISSON, M.A.: Regulation of turgor pressure in marine algae:
       ions and low-molecular-weight organic compounds. - Aust. J. Plant Physiol. 6:
       539-556, 1979.

5846 - KIRYAKOV, K., PRESOLSKA, P.: Vliyanĭe na g"stotata na poseva i toreneto v"rkhu
       nyakoi fotosintetichni pokazateli i dobiva pri samooprashenĭ linii tsarevitsa.
       [Effect of planting density and fertilizer application on some photosynthetic
       features and yield of self-pollinated maize lines.] - Rastenĭev"dni Nauki 16
       (2).: 5-12, 1979. [In Bulg, ab: E,R.]

5847 - KITTLE, D.R., GRAY, L.E.: The influence of soil temperature, moisture, porosi-
       ty, and bulk density on the pathogenicity of *Phytophthora megasperma* var.
       sojae. - Plant Dis. Rep. 63: 231-234, 1979.

5848 - KITTOCK, D.L.: Pima and upland cotton response to irrigation management. -
       Agron. J. 71: 617-619, 1979.

*5849 - KLAR, A.E., USBERTI, J.A., Jr., HENDERSON, D.W.: Differential responses of
       guinea grass populations to drought stress. - Crop Sci. 18: 853-857, 1978.

*5850 - KLEMPERER, G., EISENBACH, M., GARTY, H., CAPLAN, S.R.: The effect of salt on
       the light-induced pH changes in purple membrane from *Halobacterium halobium*. -
       In: CAPLAN, S.R., GINZBURG, M. (ed.): Energetics and Structure of Halophilic
       Microorganisms. Pp. 291-296. Elsevier / North-Holland Biomedical Press,
       Amsterdam - New York 1978.

5851 - KLENOVSKÁ, S.: Morphogenetical and physiological processes in tobacco explants
       by cultivating in water potential decreased conditions. - Acta Fac. Rerum Nat.
       Univ. Comenianae, Physiol. Plant. 16: 37-44, 1979.

5852 - KLEPPER, B., TAYLOR, H.M.: Limitations to current models describing water
       uptake by plant root systems. - In: HARLEY, J.L., SCOTT RUSSELL, R. (ed.):
       The Soil-Root Interface. Pp. 51-65. Academic Press, London - New York - San
       Francisco 1979.

*5853 - KLINGE, H., HERRERA, R.: Biomass studies in Amazon caatinga forest in southern
       Venezuela. 1. Standing crop of composite root mass in selected stands. - Trop.
       Ecol. 19: 93-110, 1978.

5854 - KLUGE, M.: The flow of carbon in Crassulacean Acid Metabolism (CAM). - In:
       GIBBS, M., LATZKO, E. (ed.): Photosynthesis II. Pp. 113-125. Springer-Verlag,
       Berlin - Heidelberg - New York 1979.

*5855 - KLUGE, M., TING, I.P.: Crassulacean Acid Metabolism. Analysis of an Ecological
       Adaptation. (Ecological studies Vol. 30.). - Springer-Verlag, Berlin - Hei-
       delberg - New York 1978.

5856 - KNITTLE, K.H., BURRIS, J.S., ERBACH, D.C.: Regression equations for rate of
soybean hypocotyl elongation by using field data. - Crop Sci. 19: 41-46, 1979.

5857 - KNYPL, J.S., JANAS, K.M.: Increasing low-temperature resistance of soybean,
Glycine max (L.) Merr., by exposure of seeds to water saturated atmosphere. -
Biol. Plant. 21: 291-297, 1979.

5858 - KOBATA, T., TAKAMI, S.: [The effects of water stress on the grain-filling in
rice.] - Jap. J. Crop Sci. 48: 75-81, 1979. [In Jap, ab: E.]

5859 - KOBAYASHI, M., SATAKE, T.: [Effective water depth for protecting rice panicles
from sterility caused by cool temperature during the booting stage.] - Jap. J.
Crop Sci. 48: 243-248, 1979. [In Jap, ab: E.]

*5860 - KOCH, K., KENNEDY, R.A.: Effect of seasonal changes in the midwest on Crassu-
lacean acid metabolism (CAM) in *Opuntia humifusa* Raf. - Plant Physiol. 61
(Suppl.): 100, 1978.

5861 - KOFOED, A.D.: The potassium cycle in cropping systems. - In: Potassium Research
- Review and Trends. Pp. 435-449. International Potash Research Institute,
Bern 1979.

*5862 - KOH, S., KUMURA, A., MURATA, Y.:[Studies on matter production in wheat plant
V. The mechanism involved in an after-effect of low night temperature.] -
Jap. J. Crop Sci. 47: 75-81, 1978. [In Jap, ab: E.]

5863 - KOHOUT, V.: Rozdíly v tvorbě hmoty plevelu a cukrovky v regulovaných bioener-
getických podmínkách. [Differences in the production of weed and sugar beet
biomass under controlled bio-energetic conditions.] - Rost. Výroba (Praha) 25:
1081- 1089, 1979. [In Czech, ab: E,G,R.]

5864 - KOĬNOV, G., VITKOV, M.: Prouchvane v"rkhu polivniya rezhim na polski grakh,
otglezhdan na opodzolen chernozem pri minimalno torene. [Study on the irrigation regime of field pea grown at a minimal fertilizer application on a podzo-
lized chernozem soil.] - Rasteniev"dni Nauki 16(1): 54-60, 1979. [In Bulg,
ab: R,E.]

*5865 - KONDO, T.: Diurnal change in leakage of electrolytes from a long-day duck-weed,
Lemna gibba G3, under osmotic stress induced by water treatment. - Plant Cell
Physiol. 19: 985-995, 1978.

5866 - KONINGS, H., JACKSON, M.B.: A relationship between rates of ethylene production by roots and the promoting or inhibiting effects of exogenous ethylene
and water on root elongation. - Z. Pflanzenphysiol. 92: 385-397, 1979.

*5867 - KOO, R.C.J.: Response of densely planted 'Hamlin' orange on two rootstocks to
low volume irrigation. - Proc. Florida State hort. Soc. 91: 8-10, 1978.

*5868 - KORANDA, J.J., CLEGG, B., STUART, M.: Radio-tracer measurement of transpira-
tion in tundra vegetation, Barrow, Alaska. - In: TIESZEN, L.L. (ed.): Vegetation and Production Ecology of an Alaskan Arctic Tundra. Pp. 359-369. Springer
-Verlag, New York - Heidelberg - Berlin 1978.

*5869 - KÖRNER, C., HOFLACHER, H., WIESER, G.: Untersuchungen zum Wasserhaushalt von
Almflächen im Gasteiner Tal. - In: CERNUSCA, A. (ed.): Ökologische Analysen
von Almflächen im Gasteiner Tal. Pp. 67-79. Universitätsverlag Wagner,
Innsbruck 1978.

*5870 - KÖRNER, C., JUSSEL, U., SCHIFFER, K.: Transpiration, Diffusionswiderstand und
Wasserpotential in verschiedenen Schichten eines Grünerlebenstandes. - In:
CERNUSCA, A. (ed.): Ökologische Analysen von Almflächen im Gasteiner Tal. Pp.
81-98. Universitätsverlag Wagner, Innsbruck 1978.

*5871 - KÖRNER, C., SCHUBERT, A.: Spaltenverhalten verschiedener Pflanzenarten auf
        Almwiesen an der zentralalpinen Waldgrenze. - In: CERNUSCA, A. (ed.): Ökolo-
        gische Analysen von Almflächen im Gasteiner Tal. Pp. 99-112. Universitäts-
        verlag Wagner, Innsbruck 1978.

5872 - KOROVIN, A.I., SEROV, V.V., YUDINA, É.V., NIKITENKO, Z.I., KRASOCHKIN, R.V.:
       Vliyanie vesennikh zamorozkov na sozrevanie i urozhaĭ yarovoĭ pshenitsy v
       zavisimosti ot temperatury vozdukha, vlazhnosti pochvy i osveshchennosti.
       [Effect of spring frosts on the maturity and yield of spring wheat as affected
       by air temperature, irradiance and soil moisture content.] - Sel'skokhoz.
       Biol. 14: 31-36, 1979. [In R, ab: E.]

5873 - KOSHUCHOWA, S., MÜNNICH, H., GÖRING, H.: Einfluss von Chlorcholinchlorid und
       Ethrel auf Zellteilung und Zellstreckung bei Primärblättern von Weizenkeim-
       pflanzen. - Biol. Plant. 21: 42-50, 1979.

5874 - KOSTOV, K.D., NAĬDENOVA, T.: Izmenenie transpiratsii i fotosinteza u molodykh
       rastenii duba krasnogo v zavisimosti ot vlazhnosti i kislotnosti pochvy.
       [Changes in young red oak transpiration and photosynthesis in relation to
       soil humidity and acidity.] - In: VAKLINOVA, S.G., VANKOVA-RADEVA, R.,
       VASILEVA, V.S. (ed.): Fotosinteticheskaya Assimilyatsiya CO_2 i Fotodykhanie.
       Pp. 160-165. Izdatel'stvo Bolgarskoĭ Akademii Nauk, Sofiya 1979. [In R.]

5875 - KOUKKARI, W.L., JOHNSON, M.A.: Oscillations of leaves of *Abutilon theophrasti*
       (velvetleaf) and their sensitivity to bentazon in relation to low and high
       humudity. - Physiol. Plant. 47: 158-162, 1979.

*5876 - KOUWENHOVEN, J.K.: Ridge quality and potato growth. - Neth. J. agr. Sci. 26:
        288-303, 1978.

5877 - KOWAL, T., KRUPIŃSKA, A.: Produktywność gatunku *Thymus pulegioides* L. w warun-
       kach naturalnych. [Productivity of the species *Thymus pulegioides* L. in natu-
       ral conditions.] - Acta agrobot. 32: 81-89, 1979. [In Pol, ab: R,E.]

5878 - KOWAL, T., PIC, S.: Produktywność gatunku *Achillea millefolium* L. w warunkach
       naturalnych. [Productivity of the species *Achillea millefolium* L. in natural
       conditions.] - Acta agrobot. 32: 91-100, 1979. [In Pol, ab: R,E.]

5879 - KOZLOWSKI, T.T.: Tree Growth and Environmental Stresses. - University of
       Washington Press, Seatle - London 1979.

5880 - KOZLOWSKI, T.T., PALLARDY, S.G.: Effects of low temperature on leaf diffusion
       resistance of *Ulmus americana* and *Fraxinus pennsylvanica* seedlings. - Can. J.
       Bot. 57: 2466-2470, 1979.

5881 - KOZLOWSKI, T.T., PALLARDY, S.G.: Stomatal responses of *Fraxinus pennsylvanica*
       seedlings during and after flooding. - Physiol. Plant. 46: 155-158, 1979.

*5882 - KRAMER, D., ANDERSON, W.P., PRESTON, J.: Transfer cells in the root epidermis
        of *Atriplex hastata* L. as a response to salinity: A comparative cytological
        and X-ray microprobe investigation. - Aust. J. Plant Physiol. 5: 739-747, 1978.

*5883 - KRAMER, P.J.: The use of controlled environments in research. - HortScience 13:
        447-451, 1978.

5884 - KRANTZ, B.A., KAMPEN, J.: Crop production systems in semi-arid tropical zones.
       - In: THORNE, D.W., THORNE, M.D. (ed.): Soil, Water and Crop Production. Pp.
       278-298. AVI Publishing Company, Inc., Westport 1979.

5885 - KRASAVTSEV, O.A.: O zaderzhke ottoka pereokhlazhdennoĭ vody iz parenkhimnykh
       kletok drevesiny yabloni. [Delay in releasing supercooled water from parenchyma
       cells of apple tree wood.] - Fiziol. Rast. 26: 415-421, 1979. [In R, ab: E.]

5886 - **KRASLOVÁ, J.**: Dynamika cukru v listech a bulvách cukrové řepy jako měřitelný příznak změn energetických podmínek. [Dynamics of sugars in the leaves and roots of sugar beet as a measurable sign of changes in energetic conditions.] - Rost. Výroba (Praha) 25: 1057-1064, 1979. [In Czech, ab: E,G,R.]

5887 - **KREMER, B.P.**: Photoassimilatory products and osmoregulation in marine *Rhodophyceae*. - Z. Pflanzenphysiol. 93: 139-147, 1979.

*5888 - **KREUZ, E.**: Neue Ergebnisse zur Ernährung und zum Wasserhaushalt des Maises - Übersichtsbeitrag. - Arch. Acker- Pflanzenbau Bodenk. 21: 327-344, 1977.

5889 - **KRISHNAMOORTHY, T.M., GOGATE, S.S., SARMA, T.P., SOMAN, S.D.**: Behaviour of tritium in succulent plants. - Ind. J. exp. Biol. 17: 401-405, 1979.

5890 - **KRISHNAMURTHY, K.V.**: Variations in the structure and ontogeny of the stomatal apparatus of the cotyledonary internode of *Commelina benghalensis* L. - Curr. Sci. 48: 117-118, 1979.

5891 - **KROCHKO, J.E., WINNER, W.E., BEWLEY, J.D.**: Respiration in relation to adenosine triphosphate content during desiccation and rehydration of a desiccation -tolerant and a desiccation-intolerant moss. - Plant Physiol. 64: 13-17, 1979.

*5892 - **KR"STEV, K.K., SAVOV, S.G.**: Fiziologo-biokhimichni izmeneniya v zabolyaloto ot vertitsilii no uvyakhvane pamukovo rastenie. [Physiological and biochemical alterations in cotton plants attacked by *Verticillium* wilt.] - Rasteniev"dni Nauki 14(3): 134-139, 1977. [In Bulg, ab: R,E.]

*5893 - **KRŮŽELA, J.**: Tvorba kořání máku setého (*Papaver somniferum* L.) při rozdílné vlhkosti pudy. [Root system formation in poppy (*Papaver somniferum* L.) at different soil moisture content.] - Rost. Výroba (Praha) 24: 193-200, 1978. [In Czech, ab: R,E,G.]

*5894 - **KRYŃSKA, W., MARTYNIAK, B.**: Wartość odżywcza kapusty wczesnej i pomidorów uprawianych na terenie falistym. [Values of early cabbage and tomatoes cultivated in a hilly area.] - Roczniki Nauk roln., Ser. A. 103(4): 79-92, 1978. [In Pol, ab: R,E.]

*5895 - **KU, S.B., EDWARDS, G.E., SMITH, D.**: Photosynthesis and nonstructural carbohydrate concentration in leaf blades of *Panicum virgatum* as affected by night temperature. - Can. J. Bot. 56: 63-68, 1978.

5896 - **KUCHEROVA, T.P., LISHCHUK, A.I., STADNIK, S.A.**: Vliyanie khlorkholinkhlorida i lateksa na vodnyĭ rezhim, soderzhanie pigmentov i bioélektricheskuyu reaktsiyu razlichnykh po zasukhoustoĭchivosti sortov abrikosa. [Effect of chlorcholine chloride and latex on water relations, content of pigments and bioelectric reaction of apricot cultivars differing in drought resistance.] - Fiziol. Biokhim. kul't. Rast. 11: 68-72, 1979. [In R, ab: E.]

5897 - **KUMAR, A., RAI, S.D.**: Irrigation and phosphorus requirements of berseem. - Ind. J. agr. Sci. 49: 73-77, 1979.

*5898 - **KUMURA, A., KOH, S., FUKAI, S., NAGANO, J.**: Diurnal variation in CO_2 exchange in stands of various field-crops. - In: MONSI, M., SAEKI, T. (ed.): Ecophysiology of Photosynthetic Productivity. JIBP Synthesis Vol. 19. Pp. 82-90. University of Tokyo Press, Tokyo 1978.

*5899 - **KÜNSTLE, E., MITSCHERLICH, G.**: Photosynthese, Transpiration und Atmung in einem Mischbestand im Schwarzwald. I. Teil: Photosynthese. - Allg. Forst- Jagdzeit. 146(3/4): 45-63, 1975.

5900 - **KURAMOTO, R.T., BREST, D.E.**: Physiological response to salinity by four salt marsh plants. - Bot. Gaz. 140: 295-298, 1979.

*5901 - KUSHNIRENKO, M.D., KORNESKU, A.C.: Metodika sravnitel'nogo opredeleniya pro-
         duktivnosti transpiratsii plodovykh rasteniĭ. [A method for determination of
         productivity of transpiration in fruit trees.] - Izv. Akad. Nauk Mold. SSR,
         Ser. biol. khim. Nauk 1975(4): 78-80, 1975. [In R.]

 5902 - KUTAS, E.N.: Vliyanie intensivnosti osveshcheniya na anatomicheskoe stroenie
         list'ev nekotorykh oranzhereĭnykh rasteniĭ. [The effect of illumination
         intensity on the anatomical structure of leaves of some greenhouse plants.]
         - Bot. Zh. 64: 1650-1657, 1979. [In R.]

*5903 - KUZNETSOVA, N.N., KARACHUN, S.I.: Sosushchaya sila korneĭ razlichnykh sortov
         yachmenya v zavisimosti ot mekhanicheskogo sostava pochvy i usloviĭ pitaniya.
         [Suction pressure in roots of different barley cultivars in dependence on
         soil mechanical properties and nutritive conditions.] - Vestnik Leningrad.
         Univ. 1978(21): 129-130, 1978. [In R.]

 5904 - LAAN, D., van der : Spatial and temporal variation in the vegetation of dune
         slacks in relation to the ground water regime. - Vegetatio 39: 43-51, 1979.

 5905 - LAKSO, A.N.: Seasonal changes in stomatal response to leaf water potential in
         apple. - J. Amer. Soc. hort. Sci. 104: 58-60, 1979.

 5906 - LAMBERT, J.R., BAKER, D.N.: Simulation of soil-root interactions. - In: HARLEY,
         J.L., SCOTT RUSSELL, R. (ed.): The Soil-Root Interface. Pp. 426-427. Academic
         Press, London - New York - San Francisco 1979.

 5907 - LAMMERS, B., STIGTER, C.J.: A battery-operated circuit for a self-timing leaf
         diffusion resistance meter. - Neth. J. agr. Sci. 27: 111-115, 1979.

 5908 - LANGE, O.L., LÖSCH, R.: Plant water relations. - Progress Bot. 41: 10-43,
         1979.

 5909 - LANGE, O.L., MEDINA, E.: Stomata of the CAM plant *Tillandsia recurvata* respond
         directly to humidity. - Oecologia 40: 357-363, 1979.

 5910 - LANGE, O.L., MEYER, A.: Mittäglicher Stomataschluss bei Aprikose (*Prunus
         armeniaca*) und Wein (*Vitis vinifera*) im Freiland trotz guter Bodenwasser
         -Versorgung. - Flora 168: 511-528, 1979.

 5911 - LANGENHEIM, J.H., STUBBLEBINE, W.H., FOSTER, C.E.: Effect of moisture stress
         on composition and yield in leaf resin of *Hymenaea courbaril*. - Biochem. Syst.
         Ecol. 7: 21-28, 1979.

 5912 - LANYI, J.K., AVRON, M., BAYLEY, S.T., BROCK, T.D., BROWN, A.D., FITT, P.S.,
         GRIFFIN, D.M., HOROWITZ, N.H., KUSHNER, D.J., LARSEN, H., NORKRANS, B.,
         TRÜPER, H.G., WEBER, J.: Life at low water activities. - In: SHILO, M. (ed.):
         Strategies of Microbial Life in Extreme Environments. Pp. 125-135. Verlag
         Chemie, Weinheim - New York 1979.

 5913 - LARKUM, A.W.D., WYN JONES, R.G.: Carbon dioxide fixation by chloroplasts iso-
         lated in glycinebetaine. A putative cytoplasmatic osmoticum. - Planta 145:
         393-394, 1979.

 5914 - LARQUÉ-SAAVEDRA, A.: Stomatal closure in response to acetylsalicylic acid
         treatment. - Z. Pflanzenphysiol. 93: 371-375, 1979.

 5915 - LARSON, D.W.: Lichen water relations under drying conditions. - New Phytol.
         82: 713-731, 1979.

 5916 - LARSON, D.W.: Preliminary studies of the physiological ecology of *Umbilicaria*
         lichens. - Can. J. Bot. 57: 1398-1406, 1979.

*5917 - LARSON, D.W., KERSHAW, K.A.: Acclimation in arctic lichens. - Nature 254:
         421-423, 1975.

5918 - LASHKEVICH, G.I.: Udobrenie, vlazhnost' pochvy, vodopotreblenie i produktivnost' yachmenya. [Fertilization, soil moisture, water consumption and barley
productivity.] - Dokl. Akad. Nauk Belorus. SSR 24(1): 79-81, 1979. [In R, ab:
E.]

5919 - LÄUCHLI, A., PFLÜGER, R.: Potassium transport through plant cell membranes
and metabolic role of potassium in plants. - In: Potassium Research - Review
and Trends. Pp. 111-163. International Potash Research Institute, Bern 1979.

5920 - LAWLOR, D.W.: Effects of water and heat stress on carbon metabolism of plants
with C_3 and C_4 photosynthesis. - In: MUSSELL, H., STAPLES, R. (ed.): Stress
Physiology in Crop Plants. Pp. 303-326. John Wiley & Sons, Inc., New York
1979.

5921 - LEACH, G.J.: Regrowth characteristics of lucerne under different systems of
grazing management. - Aust. J. agr. Res. 30: 445-465, 1979.

5922 - LEACH, J.E.: A field enclosure apparatus for measuring crop photosynthesis. -
Ann. appl. Biol. 92: 125-132, 1979.

5923 - LEACH, J.E.: Some effects of air temperature and humidity on crop and leaf
photosynthesis, transpiration and resistance to gas transfer. - Ann. appl.
Biol. 92: 287-297, 1979.

5924 - LEAVITT, J.R.C., PENNER, D.: Prevention of EPTC-induced epicuticular wax
aggregation on corn (*Zea mays*) with R-25788. - Weed Sci. 27: 47-50, 1979.

*5925 - LECHOWICZ, M.J.: Carbon dioxide exchange in *Cladina* lichens from subarctic
and temperate habitats. - Oecologia 32: 225-237, 1978.

5926 - LECHOWICZ, M.J., ADAMS, M.S.: Net CO_2 exchange in *Cladonia* lichen species
endemic to southeastern North America. - Photosynthetica 13: 155-162, 1979.

5927 - LEE, H.J., McKEE, G.W., KNIEVEL, D.P.: Determination of physiological maturity
in oat. - Agron. J. 71: 931-935, 1979.

5928 - LEGG, B.J., DAY, W., LAWLOR, D.W., PARKINSON, K.J.: The effects of drought on
barley growth: models and measurements showing the relative importance of
leaf area and photosynthetic rate. - J. agr. Sci. 92: 703-716, 1979.

*5929 - LEGGE, A.H., AMUNDSON, R.G., JAQUES, D.R., WALKER, R.B.: Field studies of
pine, spruce and aspen periodically subjected to sulfur gas emissions. -
USDA Forest Serv. Gen. Tech. Rep. NE-23: 1033-1061, 1976.

*5930 - LEIHNER, D.E., COCK, J.H.: Causes for anomalous wet-dry season yield differen-
ces in lowland rice. - Crop Sci. 17: 391-395, 1977.

5931 - LE MASSON, B., PAULIN, A.: Influence d'un déficit temporaire en eau sur le
métabolisme glucidique de l OEillet coupé (*Dianthus caryophyllus* L. var. Scania).
- C.R. Acad. Sci. Paris, Sér. D 289: 1097-1100, 1979.

*5932 - LENKA, D.: Evapotranspiration and crop co-efficient in rice. - Ind. J. Agron.
23: 351-354, 1978.

*5933 - LENKA, D., MISRA, B., SAHU, S.K., REDDY, D.S.: Effect of water management and
nitrogen application on rice. - Ind. J. Agron. 21: 420-424, 1976.

5934 - LENZ, F.: Sink-source relationships in fruit trees. - In: SCOTT, T.K. (ed.):
Plant Regulation and World Agriculture. Pp. 141-153. Plenum Press, New York
- London 1979.

*5935 - LEVITT, J.: Role of SH and SS groups in damage to biological systems at low
water activities. - In: CROWE, J.H., CLEGG, J.S. (ed.): Dry Biological Systems.
Pp. 243-256. Academic Press, New York - San Francisco - London 1978.

*5936 - LEWIS, M.C., CALLAGHAN, T.V.: Tundra. - In: MONTEITH, J.L. (ed.): Vegetation
        and the Atmosphere. Volume 2. Case Studies. Pp. 399-433. Academic Press,
        London - New York - San Francisco 1976.

5937 - LIBBERT, E.: Lehrbuch der Pflanzenphysiologie. - VEB Gustav Fischer Verlag,
       Jena 1979.

*5938 - LICHEV, B.: Poliven rezhim i vodopotreblenie na lyutsernata, otglezhdana za
        seno v raĭona na DNS-Pleven. [Water requirements and irrigation regime of
        lucerne grown for hay in the region of the Pleven state irrigation system.]
        - Rasteniev"dni Nauki 15(4): 55-62, 1978. [In Bulg, ab: R,E.]

5939 - LICHTENTHALER, H.K.: Occurence and function of prenyllipids in the photosyn-
       thetic membrane. - In: APPELQVIST, L.A., LILJENBERG, C. (ed.): Advances in
       the Biochemistry and Physiology of Plant Lipids. Pp. 57-78. Elsevier / North
       -Holland Biomedical Press, Amsterdam 1979.

5940 - LICHTENTHALER, H.K.: Effect of biocides on the development of the photosynthe-
       tic apparatus of radish seedlings grown under strong and weak light conditions.
       - Z. Naturforsch. 34C: 938-940, 1979.

*5941 - LIETH, H.: Die pflanzliche Primärproduktivität. - Bürger Univ. Informationsbl.
        Universitätsgesellschaft Osnabrück e.V. 1978(3): 38-51, 1978.

*5942 - LINACRE, E.: Swamps. - In: MONTEITH, J.L. (ed.): Vegetation and the Atmosphere.
        Volume 2. Case Studies. Pp. 329-347. Academic Press, London - New York - San
        Francisco 1976.

5943 - LINDER, S.: Photosynthesis and respiration in conifers. A classified reference
       list 1891-1977. - Stud. forest. Suec. 149: 1-71, 1979.

*5944 - LINDER, S., TROENG, E.: The seasonal course of net photosynthesis and stem
        respiration in a 20-year-old stand of Scots pine (Pinus silvestris L.). -
        In: COOMBS, J. (ed.): 4th International Congress on Photosynthesis. P. 221.
        UKISES, London 1977.

5945 - LIPE, W.N., SKINNER, J.A.: Effect of an antitranspirant on water use, growth,
       and yield of greenhouse-grown potatoes. - HortScience 14: 239-241, 1979.

*5946 - LIU, W.T., POOL, R., WENKERT, W., KRIEDEMANN, P.E.: Changes in photosynthesis,
        stomatal resistance and abscisic acid of Vitis labruscana through drought and
        irrigation cycles. - Amer. J. Enol. Viticult. 29: 239-246, 1978.

*5947 - LLOYD, D.L.: Early growth of Makarikari and Rhodes grasses in different nitro-
        gen and water regimes. - Queensland J. agr. anim. Sci. 35: 53-61, 1978.

5948 - LOCKAU, W.: The inhibition of photosynthetic electron transport in spinach
       chloroplasts by low osmolarity. - Europ. J. Biochem. 94: 365-373, 1979.

*5949 - LOMEN, D.O., WARRICK, A.W.: Time-dependent solutions to the one-dimensional
        linearized moisture flow equation with water extraction. - J. Hydrol. 39:
        59-67, 1978.

*5950 - LOMEN, D.O., WARRICK, A.W.: Linearized moisture flow with loss at the soil
        surface. - Soil Sci. Soc. Amer. J. 42: 396-400, 1978.

5951 - LONGMAN, K.A., LEAKEY, R.R.B., DENNE, M.P.: Genetic and environmental effects
       on shoot growth and xylem formation. - Ann. Bot. 44: 377-380, 1979.

5952 - LONGSTRETH, D.J., NOBEL, P.S.: Salinity effects on leaf anatomy. Consequences
       for photosynthesis. - Plant Physiol. 63: 700-703, 1979.

*5953 - LONGUENESSE, J.J.: Température nocturne et photosynthèse I. - Etude biblio-
        graphique. - Ann. agron. 29: 525-539, 1978.

5954 - LOOMIS, R.S., RABBINGE, R., NG, E.: Explanatory models in crop physiology. -
       Annu. Rev. Plant Physiol. 30: 339-367, 1979.

*5955 - LOPER, G.M., BERDEL, R.L.: Seasonal emanation of ocimene from alfalfa flowers
       with three irrigation treatments. - Crop Sci. 18: 447-452, 1978.

5956 - LORENČÍK, L., MATI, R.: Dynamika pôdnej vody pri ozimnej pšenici na Východo-
       slovenské nížine. [Dynamics of soil moisture under winter wheat grown in the
       East Slovakian lowland.] - Rost. Výroba (Praha) 25: 323-332, 1979. [In Slov,
       ab: R,E,G.]

5957 - LÖSCH, R.: Responses of stomata to environmental factors - experiments with
       isolated epidermal strips of Polypodium vulgare. II. Leaf bulk water potential,
       air humidity, and temperature. - Oecologia 39: 229-238, 1979.

5958 - LÖSCH, R.: Stomatal responses to changes in air humidity. - In: SEN, D.N.,
       CHAWAN, D.D., BANSAL, R.P. (ed.): Structure, Function and Ecology of Stomata.
       Pp. 189-216. Bishen Singh Mahendra Pal Singh, Dehra Dun 1979.

5959 - LÖSCH, R., KRUG, E.: Untersuchungen zum Stomataverhalten von Veronica persica
       und Veronica hederifolia. - In: Proc. EWRS Symp. The Influence of Different
       Factors on the Development and Control of Weeds. Pp. 89-96. 1979.

5960 - LOUÉ, A.: The interaction of potassium with other growth factors, particularly
       with other nutrients. - In: Potassium Research - Review and Trends. Pp. 407-
       433. International Potash Research Institute, Bern 1979.

5961 - LOVE, J.M., BARDEN, J.A.: The effects of ethyl 5-(4-chlorophenyl)-2H-tetrazole
       -2-acetate on net photosynthesis, stomatal resistance, and transpiration in
       'Delicious' and 'Golden Delicious' apple leaves. - HortScience 14: 515-516,
       1979.

5962 - LOVEYS, B.R., KRIEDEMANN, P.E.: The influence of abscisic acid on stomatal
       physiology. - In: SEN, D.N., CHAWAN, D.D., BANSAL, R.P. (ed.): Structure,
       Function and Ecology of Stomata. Pp. 77-101. Bishen Singh Mahendra Pal Singh,
       Dehra Dun 1979.

5963 - LÜDDERS, P., FISCHER-BÖLÜKBAŞI, T.: Einfluss von Alar und TIBA auf den Wasser-
       verbrauch unterschiedlich fruchtender Apfelbäume. - Gartenbauwissenschaft 44:
       171-177, 1979.

5964 - LÜDDERS, P., FISCHER-BÖLÜKBASI, T.: Einfluss von Alar und TIBA auf das vege-
       tative Wachstum von Apfelbäumen mit unterschiedlichem Fruchtbehang. - Garten-
       bauwissenschaft 44: 220-226, 1979.

5965 - LUDLOW, M.M., IBARAKI, K.: Stomatal control of water loss in siratro (Macro-
       ptilium atropurpureum (DC) Urb.), a tropical pasture legume. - Ann. Bot. 43:
       639-647, 1979.

*5966 - LURIE, S.: Photochemical properties of guard cell chloroplasts. - Plant Sci.
       Lett. 10: 219-223, 1977.

5967 - LURIE, S.: Photosynthetic capacity of stomatal chloroplasts and their contri-
       bution to stomatal opening. - In: SEN, D.N., CHAWAN, D.D., BANSAL, R.P. (ed.):
       Structure, Function and Ecology of Stomata. Pp. 61-76. Bishen Singh Mahendra
       Pal Singh, Dehra Dun 1979.

5968 - LUSH, W.M., RAWSON, H.M.: Effects of domestication and region of origin on
       leaf gas exchange in cowpea (Vigna unguiculata (L.) Walp.). - Photosynthetica
       13: 419-427, 1979.

5969 - LÜTTGE, U., HIGINBOTHAM, N.: Transport in Plants. - Springer-Verlag, New York
       - Heidelberg - Berlin 1979.

5970 - LYALIN, O.O., KTITOROVA, I.N., KAZARYAN, A.A., AKHMEDOV, I.S.: Elektricheskie kontakty mezhdu zamykayushcheĭ i okoloust'ichnoĭ kletkami. [Electrical contact between stomatal guard and subsidiary cells.] - Fiziol. Rast. 26: 548-551, 1979. [In R, ab: E.]

5971 - LYONS, J.M., BREIDENBACH, R.W.: Strategies for altering chilling sensitivity as a limiting factor in crop production. - In: MUSSELL, H., STAPLES, R. (ed.): Stress Physiology in Crop Plants. Pp. 180-196. John Wiley & Sons, Inc., New York 1979.

5972 - LYSHEDE, O.B.: Xeromorphic features of three stem assimilants in relation to their ecology. - Bot. J. Linn. Soc. 78: 85-98, 1979.

*5973 - MAAREL, E., van der: Experimental succession research in a coastal dune grassland, a preliminary report. - Vegetatio 38: 21-28, 1978.

*5974 - MAESTRI, M., BARROS, R.S.: Cofee. - In: ALVIM, P.de T., KOZLOWSKI, T.T. (ed.): Ecophysiology of Tropical Crops. Pp. 249-278. Academic Press, New York - San Francisco - London 1977.

5975 - MAIER-MAERCKER, U.: "Peristomatal transpiration" and stomatal movement: A controversial view. I. Additional proof of peristomatal transpiration by hygrophotography and a comprehensive discussion in the light of recent results. - Z. Pflanzenphysiol. 91: 25-43, 1979.

5976 - MAIER-MAERCKER, U.: "Peristomatal transpiration" and stomatal movement: A controversial view. II. Observation of stomatal movements under different conditions of water supply and demand. - Z. Pflanzenphysiol. 91: 157-172, 1979.

5977 - MAIER-MAERCKER, U.: "Peristomatal transpiration" and stomatal movement: A controversial view. III. Visible effects of peristomatal transpiration on the epidermis. - Z. Pflanzenphysiol. 91: 225-238, 1979.

5978 - MAIER-MAERCKER, U.: "Peristomatal transpiration" and stomatal movement: A controversial view. IV. Ion accumulation by peristomatal transpiration. - Z. Pflanzenphysiol. 91: 239-254, 1979.

*5979 - MAJAK, W., PARKINSON, P.D., WILLIAMS, R.J., LOONEY, N.E., VAN RYSWYK, A.L.: The effect of light and moisture on Columbia milkvetch toxicity in lodgepole pine forest. - J. Range Manage. 30: 423-427, 1977.

5980 - MAJUMDER, B., PRAMANIK, M.: Note on the effect of low rainfall on the flowering time in rice. - Ind. J. agr. Sci. 49: 138-139, 1979.

5981 - MAKUŠOVÁ, Z.: A quantitative study of the leaf morphology and anatomy during the ontogenetic development of Nardus stricta L. - In: RYCHNOVSKÁ, M. (ed.): Function of Grasslands in Spring Region - Kameničky Project. Pp. 121-127. Botanical Institute, Czechoslovak Academy of Sciences, Brno 1979.

5982 - MALAJCZUK, N., THEODOROU, C.: Influence of water potential on growth and cultural characteristics of Phytophthora cinnamomi. - Trans. Brit. mycol. Soc. 72: 15-18, 1979.

5983 - MALI, C.V., VARADE, S.B., MUSANDE, V.G.: Water absorption by germinating seeds of sorghum varieties at different moisture potentials. - Ind. J. agr. Sci. 49: 22-25, 1979.

5984 - MALLOCH, K.R., FENTON, R.: Inhibition of stomatal opening by analogues of abscisic acid. - J. exp. Bot. 30: 1201-1209, 1979.

5985 - MANN, H.S., KRISHNAN, A.: Crop production systems in arid tropical zones. - In: THORNE, D.W., THORNE, M.D. (ed.): Soil, Water and Crop Production. Pp. 299-317. AVI Publishing Company, Inc., Westport 1979.

5986 - MANSFIELD, T.A.: Can we control the water requirements of plants? - Trends
       biochem. Sci. 4(2): N27-N28, 1979.

5987 - MANSFIELD, T.A., WILSON, J.A.: Regulation of gas exchange in water-stressed
       plants. - In: JOHNSON, C.B. (ed.): Physiological Processes Limiting Plant
       Productivity. University of Nottingham, Sutton Bonington 1979.

5988 - MANTELL, A., MONSELISE, S.P., GOLDSCHMIDT, E.E.: Movement of tritiated water
       through young citrus plants. - J. exp. Bot. 30: 155-164, 1979.

*5989 - MAOTANI, T., NAGATA, K., KURIHARA, A., SATO, T.:[Relationship of water status
       of grapevines to drought spot.] - Bull. Fruit Tree Res. Sta. (Akitsu) E 2: 59-
       69, 1978. [In Jap, ab: E.]

*5990 - MARCZYŃSKI, S., JANKIEWICZ, L.S.: The effect of controlled temperature and
       humidity on the effectiveness of chemical defoliation of Ligustrum vulgare L.
       and Spiraea bumalda Burv. shrubs. - Acta agrobot. 31: 181-193, 1978.

*5991 - MARK, A.F., JOHNSON, P.N., WILSON, J.B.: Factors involved in the recent morta-
       lity of plants from forest and scrub along the Lake Te Anau shoreline, Fior-
       land. - Proc. N. Zeal. ecol. Soc. 24: 34-42, 1977.

5992 - MARKHART, A.H., III, FISCUS, E.L., NAYLOR, A.W., KRAMER, P.J.: Effect of
       temperature on water and ion transport in soybean and broccoli systems. -
       Plant Physiol. 64: 83-87, 1979.

5993 - MARTIN, C.K., CASSEL, D.K., KAMPRATH, E.J.: Irrigation and tillage effects on
       soybean yield in a Coastal Plain soil. - Agron. J. 71: 592-594, 1979.

5994 - MARTIN, E.S., STEVENS, R.A.: Circadian rhythms in stomatal movements. - In:
       SEN, D.N., CHAWAN, D.D., BANSAL, R.P. (ed.): Structure, Function and Ecology
       of Stomata. Pp. 251-268. Bishen Singh Mahendra Pal Singh, Dehra Dun 1979.

*5995 - MARTIN, G., MULTON, J.L.: Méthodes de références normalisées de dosage de l'eau
       dans le cas du maïs. Justification des méthodes ICC. - In: 6e Congrès Inter-
       national des Céréales et du Pain. Pp. 9-16. Winnipeg 1978.

5996 - MARTINEZ-CARRASCO, R., THORNE, G.N.: Physiological factors limiting grain
       size in wheat. - J. exp. Bot. 30: 669-679, 1979.

5997 - MARZOLA, D.L., BARTHOLOMEW, D.P.: Photosynthetic pathway and biomass energy
       production. - Science 205: 555-559, 1979.

*5998 - MATSUOKA, Y.:[Experimental studies of sulfur dioxide injury on rice plant and
       its mechanism.]- Spec. Bull. Chiba-Ken agr. exp. Sta. 7: 1-63, 1978. [In Jap,
       ab: E.]

5999 - MATTHEWS, J.M., Jr., SWINGLE, H.D.: The influence of training methods, light,
       and moisture on acid content of tomato fruits. - HortScience 14: 262-263,
       1979.

*6000 - MAUER, J., MAYO, J.M., DENFORD, K.: Comparative ecophysiology of the chromoso-
       me races in Viola adunca J.E.Smith. - Oecologia 35: 91-104, 1978.

*6001 - McDANIEL, G.L., BRESENHAM, G.L.: Use of antitranspirants to improve water
       relations of cineraria. - HortScience 13: 466-467, 1978.

*6002 - McDOLE, R.E., McMASTER, G.M.: Effects of moisture stress and nitrogen ferti-
       lization on tuber nitrate-nitrogen content. - Amer. Potato J. 55: 611-619,
       1978.

6003 - McKEEVER, J., FERGUSON, J.A., MORELOCK, T.E.: A technique for timing drip
       irrigation of staked tomatoes. - Arkansas Farm Res. 28(3): 7, 1979.

6004 - McKINION, J.M., WEAVER, R.E.C.: Simulation of plant response to primary stress vectors. - Trans. ASAE 22: 586-591, 597, 1979.

6005 - McLAUGHLIN, S.B., SHRINER, D.S., McCONATHY, R.K., MANN, L.K.: Effects of SO_2 dosage kinetics and exposure frequency on photosynthesis and transpiration of kidney beans (*Phaseolus vulgaris* L.). - Environ. exp. Bot. 19: 179-191, 1979.

*6006 - McNEE, P., WARRELL, L.A., MUYZENBERG, E.W.B., van den: Influence of water stress on yield and quality of flue-cured tobacco. - Aust. J. exp. Agr. anim. Husb. 18: 726-731, 1978.

*6007 - McWHORTER, C.G., AZLIN, W.R.: Effects of environment on the toxicity of gly-phosate to johnsongrass (*Sorghum halepense*) and soybeans (*Glycine max*). - Weed Sci. 26: 605-608, 1978.

*6008 - MEDINA, E., SOBRADO, M., HERRERA, R.: Significance of leaf orientation for leaf temperature in an Amazonian sclerophyll vegetation. - Radiat. Environ. Biophys. 15: 131-140, 1978.

6009 - MEIDNER, H.: Water uptake and translocation, stomatal movements and transpi-ration. - In: PERRY, R.A., GOODALL, D.W. (ed.): Arid-land Ecosystems: Structu-re, Functioning and Management. Volume 1. Pp. 491-507. Cambridge University Press, Cambridge 1979.

6010 - MEIDNER, H., BANNISTER, P.: Pressure and solute potentials in stomatal cells of *Tradescantia virginiana*. - J. exp. Bot. 30: 255-265, 1979.

*6011 - MEKHANDZHIEVA, A., KOLEVA, S.: V"rkhu posloĭnoto izpolzuvane na vodata ot pochvata. [On the withdrawal of water from the soil layers.] - Rasteniev"dni Nauki 15(4): 46-54, 1978. [In Bulg, ab: R,E.]

6012 - MELCAREK, P.K., BROWN, G.N.: Chlorophyll fluorescence monitoring of freezing point exotherms in leaves. - Cryobiology 26: 69-73, 1979.

*6013 - MENGEL, K., KIRKBY, E.A.: Principles of Plant Nutrition. - International Potash Research Institute, Bern 1978.

*6014 - MENOUX-BOYER, Y., De PARCEVAUX, S.: Quelques effets et post-effets de séche-resses sur la croissance des plantes: cas particulier du Lin textile. - Bull. Soc. ecophysiol. 3: 54-55, 1978.

*6015 - MEON, S., WALLACE, H.R., FISHER, J.M.: Water relations of tomato (*Lycopersicon esculentum* Mill. cv. Early Dwarf Red) infected with *Meloidogyne javanica* (Treub), Chitwood. - Physiol. Plant Pathol. 13: 275-281, 1978.

6016 - MERRILL, S.D., RAWLINS, S.L.: Distribution and growth of sorghum roots in response to irrigation frequency. - Agron. J. 71: 738-745, 1979.

*6017 - MERT, H.H., DĪZBAY, M.: The effect of osmotic pressure and salinity of the medium on the growth and sporulation of *Aspergillus niger* and *Paecilomyces lilacinum* species. - Mycopathologia 61: 125-127, 1977.

*6018 - MEYER, W.S., ALSTON, A.M.: Wheat responses to seminal root geometry and sub-soil water. - Agron. J. 70: 981-986, 1978.

*6019 - MEYER, W.S., GREACEN, E.L., ALSTON, A.M.: Resistance to water flow in the seminal roots of wheat. - J. exp. Bot. 29: 1451-1461, 1978.

6020 - MICHALCZYK, K.-W.: The effect of climatic variations on the significance of agricultural planning data. - Agr. Meteorol. 20: 319-326, 1979.

6021 - MICHEL, B.E.: Correction of thermal gradient errors in stem thermocouple hygrometers. - Plant Physiol. 63: 221-224, 1979.

6022 - MIDDLETON, J.E., PROEBSTING, E.L., ROBERTS, S.: Apple orchard irrigation by trickle and sprinkler. - Trans. ASAE 22: 582-584, 1979.

*6023 - MIKHAÏLOVA, A.V.: Nekotorye osobennosti raboty fotosinteticheskogo apparata list'ev yachmenya pod vliyaniem zatopleniya pochvy vodoÏ i vitamina PP. [Certain characteristics of activity of the photosynthetic apparatus of barley leaves under the effect of soil flooding and vitamin PP.] - Nauch. Dokl. vyssh. Shkoly, biol. Nauki 20(9): 104-108, 1977. [In R.]

6024 - MILBORROW, B.V.: Antitranspirants and the regulation of abscisic acid content. - Aust. J. Plant Physiol. 6: 249-254, 1979.

6025 - MILBURN, J.A.: Water flow in Plants. - Longman, London - New York 1979.

6026 - MILLER, P.C., POOLE, D.K.: Patterns of water use by shrubs in southern California. - Forest Sci. 25: 84-98, 1979.

*6027 - MILLER, P.C., STONER, W.A., EHLERINGER, J.R.: Some aspects of water relations of arctic and alpine regions. - In: TIESZEN, L.L. (ed.): Vegetation and Production Ecology of an Alaskan Arctic Tundra. Pp. 343-357. Springer-Verlag, New York - Heidelberg - Berlin 1978.

*6028 - MILLER, R.W.: Osmotically induced removal of water from fungal cells as determined by a spin probe technique. - Plant Physiol. 62: 741-745, 1978.

6029 - MILLS, J.D., HIND, G.: Light-induced Mg^{2+} ATPase activity of coupling factor in intact chloroplasts. - Biochim. biophys. Acta 547: 455-462, 1979.

6030 - MILLS, J.T., WALLACE, H.A.H.: Microflora and condition of cereal seeds after a wet harvest. - Can. J. Plant Sci. 59: 645-651, 1979.

6031 - MILNE, R.: Water loss and canopy resistance of a young Sitka spruce plantation. - Boundary-Layer Meteorol. 16: 67-81, 1979.

6032 - MILTHORPE, F.L., MOORBY, J.: An Introduction to Crop Physiology. Second Edition. - Cambridge University Press, Cambridge - London - New York - New Rochelle - Melbourne - Sydney 1979.

6033 - MILTHORPE, F.L., THORPE, N., WILLMER, C.M.: Stomatal metabolism - A current assessment of its features in *Commelina*. - In: SEN, D.N., CHAWAN, D.D., BANSAL, R.P. (ed.): Structure, Function and Ecology of Stomata. Pp. 121-142. Bishen Singh Mahendra Pal Singh, Dehra Dun 1979.

6034 - MIRHADI, M.J., KOBAYASHI, Y.: Studies on the productivity of grain sorghum. II. Effects of wilting treatments at different stages of growth on the development, nitrogen uptake and yield of irrigated grain sorghum. - Jap. J. Crop Sci. 48: 531-542, 1979.

*6035 - MIROSHNICHENKO, Yu.M.: Izmenenie struktury saksaul'nikov pod vliyaniem vypasa i zarastaniya mkhom v Vostochnykh Karakumakh. [Influence of grazing and moss -growing on the structure of *Haloxylon* stands in Eastern Karakum.]- Problemy Osvoeniya Pustyn' (Ashkhabat) 5: 18-28, 1975. [In R, ab: E,Turkm.]

*6036 - MISAGHI, I.J., GROGAN, R.G., DUNIWAY, J.M., KIMBLE, K.A.: Influence of environment and culture media on spore morphology of *Alternaria alternata*. - Phytopathology 68: 29-34, 1978.

*6037 - MISHKIND, M., PALEVITZ, B.A.: Guard cell differentiation in *Cyperus esculentus*. - Plant Physiol. 61(Suppl.): 116, 1978.

6038 - MITSCH, W.J., EWEL, K.C.: Comparative biomass and growth of cypress in Florida Wetlands. - Amer. Midland Natur. 101: 417-426, 1979.

6039 - MOAWAD, M.: Ecophysiology of vesicular-arbuscular mycorrhiza in the tropics. -
In: HARLEY, J.L., SCOTT RUSSELL, R. (ed.): The Soil-Root Interface. Pp. 197-
209. Academic Press, London - New York - San Francisco 1979.

*6040 - MOCA, V.: Cercetări privind influența lucrărilor de drenaj asupra solurilor ou
exces de umiditate din Depresiunea Baia-Sasca. [Researches regarding the
influence of draining works on soils with humidity excess in Baia-Sasca De-
pression.] - Lucrări Stiinț. Inst. agr. "Ion Ionescu de la Brad" Iași, Ser.
Agr. 1977: 9-12, 1977. [In Roum, ab: R,E.]

6041 - MOLDAU, H.: Versatility of stomatal control. - In: SEN, D.N., CHAWAN, D.D.,
BANSAL, R.P. (ed.): Structure, Function and Ecology of Stomata. Pp. 175-188.
Bishen Singh Mahendra Pal Singh, Dehra Dun 1979.

*6042 - MØLLER, I.M.: Balance between cyanide-sensitive and -insensitive respiration
in wheat root mitochondria, as influenced by salt concentration in the plant
growth medium. - Physiol. Plant. 42: 157-162, 1978.

6043 - MOLZ, F.J., KERNS, D.V., Jr., PETERSON, C.M., DANE, J.H.: A circuit analog
model for studying quantitative water relations of plant tissues. - Plant
Physiol. 64: 712-716, 1979.

6044 - MOMEN, N.N., CARLSON, R.E., SHAW, R.H., ARJMAND, O.: Moisture-stress effects
on the yield components of soybean cultivars. - Agron. J. 71: 86-90, 1979.

*6045 - MONTEITH, J.L. (ed.): Vegetation and the Atmosphere. Volume 1. Principles. -
Academic Press, London 1975.

*6046 - MONTEITH, J.L. (ed.): Vegetation and the Atmosphere. Volume 2. Case Studies. -
Academic Press, London - New York - San Francisco 1976.

*6047 - MONTEITH, J.L.: Climate. - In: ALVIM, P.de T., KOZLOWSKI, T.T. (ed.): Ecophy-
siology of Tropical Crops. Pp. 1-27. Academic Press, New York - San Francisco
- London 1977.

*6048 - MONTEITH, J.L.: Climate and efficiency of crop production in Britain. - Phil.
Trans. roy. Soc. London B 281: 277-294, 1977.

*6049 - MOONEY, H.A., BJÖRKMAN, O., BERRY, J.: Photosynthetic adaptations to high
temperature. - In: HADLEY, N.F. (ed.): Environmental Physiology of Desert
Organisms. Pp. 138-151. Dowden, Hutchinson and Ross, Inc., Stroudsburg 1975.

6050 - MOORE, T.C.: Biochemistry and Physiology of Plant Hormones. - Springer-Verlag,
New York - Heidelberg - Berlin 1979.

6051 - MOOSAVI-NIA, H., DORE, J.: Factors affecting glyphosate activity in *Imperata
cylindrica* (L) Beau. and *Cyperus rotundus* L. I. Effect of soil moisture. -
Weed Res. 19: 137-143, 1979.

*6052 - MORAES, V.H.F.: Rubber. - In: ALVIM, P.de T., KOZLOWSKI, T.T. (ed.): Ecophy-
siology of Tropical Crops. Pp. 315-331. Academic Press, New York - San Francisco - London 1977.

*6053 - MOROHASHI, Y.: Development of respiratory metabolism in seeds during hydration.
- In: CROWE, J.H., CLEGG, J.S. (ed.): Dry Biological Systems. Pp. 225-240.
Academic Press, New York - San Francisco - London 1978.

*6054 - MOROZOV, V.L., BELAYA, G.A.: Akkumulyatsiya solnechnoĭ energii kamchatskim
krupnotrav'em v razlichnykh ėkologicheskikh usloviyakh. [Accumulation of solar
energy by tall herbaceous vegetation of the Kamchatka peninsula in different
ecological conditions.] - Ėkologiya 1978(1): 34-41, 1978. [In R.]

6055 - MORRISON BAIRD, L.A., LEOPOLD, A.C., BRAMLAGE, W.J., WEBSTER, B.D.: Ultrastructural modifications associated with imbibition of the soybean radicle.-
Bot. Gaz. 140: 371-377, 1979.

6056 - MORROW, L.A., POWER, J.F.: Effect of soil temperature on development of pe-
       rennial forage grasses. - Agron. J. 71: 7-10, 1979.

*6057 - MOTT, G.O., POPENOE, H.L.: Grasslands. - In: ALVIM, P.de T., KOZLOWSKI, T.T.
       (ed.): Ecophysiology of Tropical Crops. Pp. 156-186. Academic Press, New York
       - San Francisco - London 1977.

6058 - MOUTONNET, P.: Un système d'irrigation automatique des cultures: principe et
       simulation theorique. - Agr. Meteorol. 20: 25-39, 1979.

6059 - MOUTONNET, P., COUCHAT, P.: Mesure journalière sur cycle végétatif complet
       des échanges gazeux: photosynthèse, transpiration et respiration nocturne
       d'une culture de Maïs conduite sur colonnes de sol. - Physiol. Plant. 47:
       39-43, 1979.

6060 - MOVCHAN, L.T.: Vodoobmen izolirovannykh list'ev i regulyatory rosta. [Water
       exchange of isolated leaves and growth regulators.] - Fiziol. Biokhim. kul't.
       Rast. 11: 258-261, 1979. [In R, ab: E.]

6061 - MOYER, J.R.: Soil organic matter, moisture, and temperature: Effect on wild
       oats control with trifluralin. - Can. J. Plant Sci. 59: 763-768, 1979.

6062 - MOZHAEVA, L.V., PIL'SHCHIKOVA, N.V., KUZINA, V.I.: Izuchenie prirody dvizhush-
       chei sily placha rastenii s ispol'zovaniem khimicheskikh vozdeistvii. [Study
       of character of motive forces of plant exudation by use of chemicals.] - Izv.
       TSChA, Ser. Fiziol. Rast. Mikrobiol. 1979(1): 3-9, 1979. [In R, ab: E.]

*6063 - MRÁZ, K.: Změny vodního režimu illimerizované podzolové půdy chlumního smrko-
       vého porostu v letech 1973 az 1975. [Changes in water regime of illimerized
       podzol soil of spruce stand in the course of 1973 - 1975.] - Lesnictví (Praha)
       24: 101-118, 1978. [In Czech, ab: R,E,G,F.]

6064 - MRÁZ, K.: Vliv dokončených vodohospodářských úprav na lužní lesy jižní Moravy.
       [Riverine forests of south Moravia as affected by performend water-management
       regulations.] - Lesnictví (Praha) 25: 45-56, 1979. [In Czech, ab: R,E,G,F.]

6065 - MRÁZ, K.: Produkční význam půdní vlhkosti v chlumních smrkových porostech.
       [Production importance of soil moisture in Norway spruce forest stands in
       hills.] - Práce VÚLHM (Jíloviště - Strnady) 55: 7-20, 1979. [In Czech, ab: R,E.]

6066 - MULDER, C.E.G., NALEWAJA, J.D.: Influence of moisture on soil-incorporated
       diclofop. - Weed Sci. 27: 83-87, 1979.

6067 - MÜLLER, F., SCHARF, H.: Der Einfluss der Vorsaatjarowisation auf den Trocken-
       massegehalt und auf den Gehalt an gebundenen Wasser während des Winters bei
       Gerste (Hordeum vulgare L.). - Arch. Züchtungsforsch. 9: 101-108, 1979.

6068 - MULLER, W.H.: Botany: A Functional Approach. Fourth Edition. - Macmillan
       Publishing Co., Inc., New York 1979, Collier Macmillan Publishers, London 1979.

6069 - MUNNS, R., BRADY, C.J., BARLOW, E.W.R.: Solute accumulation in the apex and
       leaves of wheat during water stress. - Aust. J. Plant Physiol. 6: 379-389,
       1979.

6070 - MURÍN, A.: Effects of high osmotic potential of a medium on mitotic cycle in
       roots of Vicia faba L. - Biol. Plant. 21: 345-350, 1979.

6071 - MURPHY, J.W., NELSON, S.H.: Effects of turf on percolation and water holding
       capacity of three sand mixtures. - N. Zeal. J. exp. Agr. 7: 245-248, 1979.

*6072 - MURRAY, D.B.: Coconut palm. - In: ALVIM, P.de T., KOZLOWSKI, T.T. (ed.):
       Ecophysiology of Tropical Crops. Pp. 384-407. Academic Press, New York - San
       Francisco - London 1977.

6073 - MURTHY, G.S.R., INAMDAR, J.A.: Morphogenetic effects of various growth substances on the cotyledonary stomata of brinjal and tomato. - Biol. Plant. 21:
328-335, 1979.

6074 - NAGARAJAH, S.: Differences in cuticular resistance in relation to transpiration in tea (*Camellia sinensis*). - Physiol. Plant. 46: 89-92, 1979.

6075 - NAGARAJAH, S.: The effect of potassium deficiency on stomatal and cuticular
resistance in tea (*Camellia sinensis*). - Physiol. Plant. 47: 91-94, 1979.

*6076 - NAIM, R., NEUMANN, P.M.: Mechanism of senescence induction by silicone oil. -
Physiol. Plant. 42: 57-60, 1978.

*6077 - NAKAGAWA, Y.: Occurrence and prevention of drought injury to crops in western
Japan. - In: TAKAHASHI, K., YOSHINO, M.M. (ed.): Climatic Change and Food
Production. Pp. 231-236. University of Tokyo Press, Tokyo 1978.

*6078 - NAKAMO, K.: [Changes in soil physical properties of clayey soil by conversion
from ill-drained paddy field into upland field.] - Bull. Hokuriku nat. agr.
exp. Sta. 21: 63-94, 1978. [In Jap, ab: E.]

*6079 - NAKAMURA, H., HASHIMOTO, S., OTA, Y., NAKANO, M.: [Photochemical oxidants
injury in rice plants. I. Occurrence of photochemical oxidants injury in rice
plants at Kanto area and its symptoms.] - Proc. Crop Sci. Soc. Jap. 44: 312
-319, 1975. [In Jap, ab: E.]

*6080 - NAKAMURA, H., OTA, Y.: An injury to rice plants caused by photochemical oxi-
dants in Japan. - JARQ 12(2): 69-73, 1978.

*6081 - NAKAZAWA, S.: Localized plasmolysis in *Fucus* eggs as a new pattern of group
effect. - Bull. Yamagata Univ., Natur. Sci. 9: 255-257, 1977.

6082 - NANDA, H.P., KAR, R.K., KABI, T.: Symptomological changes through seasonal
cycle in mosaic virus infected papaya. - Geobios 6: 235-237, 1979.

6083 - NASSAR, N.M.A.: Three Brazilian *Manihot* species with tolerance to stress conditions. - Can. J. Plant Sci. 59: 553-555, 1979.

6084 - NASSERY, H., OGATA, G., MAAS, E.V.: Sensitivity of sesame to various salts. -
Agron. J. 71: 595-597, 1979.

*6085 - NÁTR, L.: Programování, modelování a prognózy výnosu obilovin. [Programming,
modelling and prognoses of cereal yields.]- Stud. Inform. ÚVTIZ, Základní
Vědy Zeměd. 1978(2): 1-84, 1978. [In Czech, ab: R,E.]

6086 - NAUMANN, W.-D., PLANCHER, B.: Untersuchungen zur klimatisierenden Beregnung
von Obstgehölzen. II. Nettoassimilation bei Apfelsämlingen als Modellpflanzen
und bei Jungbäumen von "Golden Delicious". - Gartenbauwissenschaft 44: 22-26,
1979.

6087 - NAVARA, J.: Die Veränderungen des Wassergehaltes in den Blättern der Aprikose
während der Vegetationsperiode. - Biológia (Bratislava) 34: 273-282, 1979.

6088 - NAVARA, J.: Der Verlauf der Transpirationsintensität der Blätter von *Prunus
armeniaca* L. während der Vegetationsperiode. - Biológia (Bratislava) 34:
523-529, 1979.

*6089 - NEGISI, K.: Respiration in forest trees. - In: SHIDEI, T., KIRA, T. (ed.):
Primary Productivity of Japanese Forests - Productivity of Terrestrial Communities. Pp. 86-99. University of Tokyo Press, Tokyo 1977.

*6090 - NEMCHENKO, É.P., NOVIKOV, V.S.: Anatomicheskoe stroenie list'ev nekotorykh
vidov *Lilium* L. [Anatomical structure of leaves of some *Lilium* L. species.] -
Biol. Nauki (Moskva) 1978(1): 90-97, 1978. [In R.]

6091 - NEMCHENKO, È.P., NOVIKOV, V.S.: Anatomicheskoe stroenie list'ev nekotorykh
       vidov *Lilium* L. II. [Anatomical structure of leaves of some *Lilium* L. species.
       II.] - Biol. Nauki (Moskva) 1979 (6): 53-58, 1979. [In R.]

6092 - NESTEROVICH, N.D., LUCHKOV, A.I., DERYUGINA, T.F.: Izmenenie nekotorykh anato-
       micheskikh pokazateleĭ khvoi dvuletnikh seyantsev sosny obyknovennoĭ, vyra-
       shchennykh pri raznykh urovnyakh gruntovykh vod. [Changes of some anatomical
       features of two-year old *Pinus silvestris* L. seedlings grown at different
       groundwater levels.] - Dokl. Akad. Nauk Beloruss. SSR 23: 646-648, 671, 1979.
       [In R, ab: E.]

*6093 - NEYRA, C.A., HAGEMAN, R.H.: Relationships between carbon dioxide, malate, and
       nitrate accumulation and reduction in corn (*Zea mays* L.) seedlings. - Plant
       Physiol. 58: 726-730, 1976.

6094 - NICHOLAIDES, J.III.: Crop production systems on acid soils in humid tropical
       America. - In: THORNE, D.W., THORNE, M.D. (ed.): Soil, Water and Crop Produc-
       tion. Pp. 243-277. AVI Publishing Company, Inc., Westport 1979.

6095 - NICHOLLS, P.B.: Induction of sensitivity to gibberellic acid in developing
       wheat caryopses: effect of rate of desiccation. - Aust. J. Plant Physiol. 6:
       229-240, 1979.

*6096 - NICKELL, L.G.: Sugarcane. - In: ALVIM, P. de T., KOZLOWSKI, T.T. (ed.): Eco-
       physiology of Tropical Crops. Pp. 89-111. Academic Press, New York - San
       Francisco - London 1977.

*6097 - NIENHUIS, P.H.: Dynamics of benthic algal vegetation and environment in Dutch
       estuarine salt marshes, studied by means of permanent quadrats. - Vegetatio
       38: 103-112, 1978.

6098 - NIIMI, Y., TORIKATA, H.: Changes in photosynthesis and respiration during
       berry development in relation to the ripening of Delaware grapes. - J. Jap.
       Soc. hort. Sci. 47: 448-453, 1979.

*6099 - NIKOLOV, G.: Arkhitektonika na pamukoviya posev v zavisimost ot vodoobezpe-
       chenostta mu. [Architecture of cotton crop as dependent on its water supply.] -
       Rasteniev"dni Nauki 15(9-10): 97-107, 1978. [In Bulg, ab: R, E.]

6100 - NINOVA, D., ZHEKOV, Zh.Ĭ., DANAILOVA, S.: Anatomo-morfologichni prouchvaniya
       na tsarevitsa, otglezhdana pri razlichna pochvena vlazhnost. [Anatomical and
       morphological studies on maize grown at varying soil moisture content.] -
       Rasteniev"dni Nauki 16(6): 16-26, 1979. [In Bulg, ab: R, E.]

6101 - NITSCH, C., GODARD, M.: The role of hormones in promoting and developing
       growth to select new varieties in sterile culture. - In: SCOTT, T.K. (ed.):
       Plant Regulation and World Agriculture. Pp. 49-62. Plenum Press, New York -
       London 1979.

*6102 - NJØS, A.: Effects of tractor trafic and liming on yields and soil physical
       properties of a silty clay loam soil. - Meld. Norges Landbrukshøgsk. 57(24):
       1-26, 1978.

6103 - NOBEL, P.S., HARTSOCK, T.L.: Environmental influences on open stomateś of a
       crassulacean acid metabolism plant, *Agave deserti*. - Plant Physiol. 63: 63-66,
       1979.

6104 - NORTHOLT, M.D., VAN EGMOND, H.P., PAULSCH, W.E.: Ochratoxin A production by
       some fungal species in relation to water activity and temperature. - J. Food
       Protection 42: 485-490, 1979.

6105 - NOWAKOWSKI, W.: Wpływ IAA na aktywnosc oksydazy kwasu indolilooctowego w siew-
       kach pszenicy ozimej i kukurydzy w warunkach stresu osmotycznego. [Influence
       of IAA on IAA-oxidase activity in winter wheat and maize seedlings under condi-
       tions of osmotic stress.] - Acta agrobot. 32 (1): 101-107, 1979. [In Pol, ab: E.]

6106 - NUTALL, W.F., ZANDSTRA, H.G., BOWREN, K.E.: Yield and N percentage of spring wheat as affected by phosphate fertilizer, moisture use, and available soil P and N. - Agron. J. 71: 385-391, 1979.

6107 - NYÉKI, J., IFJÚ, Z., GERGELY, I.: Effect of irrigation on flowering in the sour cherry variety "Érdi bötermö". - Acta agron. Acad. Sci. Hung. 28: 59-65, 1979.

6108 - OBERBAUER, S., MILLER, P.C.: Plant water relations in montane and tussock tundra vegetation types in Alaska. - Arctic Alpine Res. 11: 69-81, 1979.

*6109 - OECHEL, W.C., SVEINBJÖRNSSON, B.: Primary production processes in arctic bryophytes at Barrow, Alaska. - In: TIESZEN, L.L. (ed.): Vegetation and Production Ecology of an Alaskan Arctic Tundra. Pp. 269-298. Springer-Verlag, New York - Heidelberg - Berlin 1978.

6110 - OGAWA, T.: Stomatal responses to light and CO_2 in greening wheat leaves. - Plant Cell Physiol. 20: 445-452, 1979.

6111 - O'LEARY, J.W.: Yield potential of halophytes and xerophytes. - In: GOODIN, J. R., NORTHINGTON, D.K. (ed.): Arid Land Plant Resources. Pp. 574-581. Texas Tech. University, Lubbock 1979.

*6112 - OLECH, K., STANEK, R.: Wpływ okresowego zaciemniania siewek fasoli na intensywność transpiracji jako wskaźnika stopnia otwierania się aparatów szparkowych. [Influence of periodical darkening of bean seedlings on the intensity of transpiration as an index of degree of openning of the stomata apparatus.] - Ann. Univ. Mariae Curie-Skłodowska, Sect. E. Agr. 30: 203-212, 1975. [In Pol, ab: E,R.]

6113 - OLSZTA, W.: Some effects of soil water conditions on the root growth of grass in the peat muck soil. - In: HARLEY, J.L., SCOTT RUSSELL, R. (ed.): The Soil--Root Interface. Pp. 430-431. Academic Press, London - New york - San Francisco 1979.

*6114 - OMBRELLO, T.M., GARRISON, S.A.: Establishing asparagus from seedling transplants. - HortScience 13: 663-664, 1978.

6115 - ORAM, P.A.: Crop production systems in the arid and semi-arid warm temperate and Mediterranean zones. - In: THORNE, D.W., THORNE, M.D. (ed.): Soil, Water and Crop Production. Pp. 193-228. AVI Publishing Company, Inc., Westport 1979.

6116 - ORMROD, D.P.: Night temperature and water effects on bean response to supplementary ammonium nitrogen. - HortScience 14: 637-639, 1979.

6117 - OSMOND, C.B., LUDLOW, M.M., DAWIS, R., COWAN, I.R., POWLES, S.B., WINTER, K.: Stomatal responses to humidity in *Opuntia inermis* in relation to control of CO_2 and H_2O exchange patterns. - Oecologia 41: 65-76, 1979.

6118 - OSMOND, C.B., NOTT, D.L., FIRTH, P.M.: Carbon assimilation patterns and growth of the introduced CAM plant *Opuntia inermis* in Eastern Australia. - Oecologia 40: 331-350, 1979.

6119 - O'SULLIVAN, J.: Response of peppers to irrigation and nitrogen. - Can. J. Plant Sci. 59: 1085-1091, 1979.

6120 - O'TOOLE, J.C., CRUZ, R.T., SEIBER, J.N.: Epicuticular wax and cuticular resistance in rice. - Physiol. Plant. 47: 239-244, 1979.

6121 - O'TOOLE, J.C., CRUZ, R.T., SINGH, T.N.: Leaf rolling and transpiration. - Plant Sci. Lett. 16: 111-114, 1979.

*6122 - O'TOOLE, J.C., MOYA, T.B.: Genotypic variation in maintenance of leaf water potential in rice. - Crop Sci. 18: 873-876, 1978.

6123 - OUTLAW, W.H.,Jr., MANCHESTER, J.: Guard cell starch concentration quantitat-
ively related to stomatal aperture. - Plant Physiol. 64: 79-82, 1979.

6124 - OUTLAW, W.H.,Jr., MANCHESTER, J., DiCAMELLI, C.A.: Histochemical approach to
properties of *Vicia faba* guard cell phosphoenolpyruvate carboxylase. - Plant
Physiol. 64: 269-272, 1979.

*6125 - OVERDIECK, D.: CO_2-Gaswechsel und Transpiration von Sonnen- und Schattenblättern
bei unterschiedlichen Strahlungsqualitäten. - Ber. Deut. bot. Ges. 91: 633-
644, 1978.

*6126 - PADILLA, C., SARROCA, J., FEBLES, G.: Effects of sowing depth and humidity in
two Cuban soils on the germination of *Panicum maximum* Jacq. - Cuban J. agr.
Sci. 12: 317-323, 1978.

*6127 - PALATINUS, L.: Výskyt steblolamu (*Cercosporella herpotrichoides* Fron.) a čer-
nania päty stebla (*Ophiobolus graminis* Sacc.) na ozimnej pšenici pri závlahe.
[Incidence of *Cercosporella herpotrichoides* and *Ophiobolus graminis* in irrigated winter wheat.]- Rost. Výroba (Praha) 24: 1207-1215, 1978. [In Slov, ab:
R,E,G.]

*6128 - PALFI, G., NAZEZ, R., K"DREV, T.: Vliyanie na nyakoi rastezhni veshchestva i
KCl v"rkhu s"d"rzhanieto na prolin i svobodni aminokiselini pri voden defitsit.
[The effect of some growth substances and KCl on proline and free amino acid
content during water stress.] - Fiziol. Rast. (Sofia) 2(3): 10-18, 1976. [In
Bulg, ab: R,E.]

6129 - PALLAND, C.L.: The "evaporation brush", an evaporimeter for measuring the potential evaporation of meadow grass. - J. Hydrol. 41: 363-369, 1979.

6130 - PALLARDY, S.G., KOZLOWSKI, T.T.: Relationships of leaf diffusion resistance of
Populus clones to leaf water potential and environment. - Oecologia 40: 371-
380, 1979.

6131 - PALLARDY, S.G., KOZLOWSKI, T.T.: Stomatal response of *Populus* clones to light
intensity and vapor pressure deficit. - Plant Physiol. 64: 112-114, 1979.

6132 - PALLAS, J.E.,Jr., STANSELL, J.R., KOSKE, T.J.: Effects of drought on Florunner
peanuts. - Agron. J. 71: 853-858, 1979.

*6133 - PALTA, J.P., LEVITT, J., STADELMANN, E.J.: Freezing tolerance of onion bulbs
and significance of freeze-induced tissue infiltration. - Cryobiology 14:
614-619, 1977.

6134 - PĂLTINEANU, R.: Studiul evapotranspiraţiei la floarea-soarelui ca element
principal în prognoza irigării. [Evapotranspiration as principal element for
scheduling irrigation needs of sunflower.] - An. Inst. Cercetări Pentru Cereale
Plante tehnice Fundulea 44: 329-338, 1979. [In Roum, ab: R,E.]

*6135 - PĂLTINEANU, R., SIPOŞ, G.: Utilizarea evapotranspiraţiei în prognoza udărilor
la grîu si orz. [Use of evapotranspiration in forecasts of the irrigation
applied to wheat and barley.] - An. Inst. Cercetări Pentru Cereale Plante
tehnice Fundulea 43: 323-334, 1978. [In Roum, ab: R,E.]

6136 - PALZKILL, D.A., TIBBITTS, T.W., WILLIAMS, P.H.: Enhancement of calcium transport to inner leaves of cabbage for prevention of tipburn. - J. Amer. Soc.
hort. Sci. 101: 645-648, 1979.

*6137 - PAPÁNEK, D.: Vplyv niektorých fungicídov na veľkosť a otvorenosť prieduchov
na viniči. [Influence of some fungicides on stomata size and opening in *Vitis
vinifera*.] - Vinohrad (Bratislava) 16: 225-227, 1978. [In Slov.]

*6138 - PARGNEY, J.C., THALOUARN, P.: Etude cytologique de l'embryon du Pin d'Alep
lors de la germination: influence d'un traitement stimulant par administration
successive d'ions mercure et chlore. - Can. J. Bot. 56: 2931-2936, 1978.

6139 - PÂRJOL, L., POPA, F.G., HURDUC, N.: Cercetări privind rezistenţa la secetă a
unor soiuri şi linii de fasole. [Researches regarding drought resistance of
some bean cultivars and lines.] - An. Inst. Cercetări Pentru Cereale Plante
tehnice Fundulea 44: 415-426, 1979. [In Roum, ab: R,E.]

6140 - PARKER, J.: Effects of defoliation and root height above a water table on some
red oak root metabolites. - J. Amer. Soc. hort. Sci. 104: 417-421, 1979.

6141 - PARKINSON, K.J., DAY, W.: The use of orifices to control the flow rate of
gases. - J. appl. Ecol. 16: 623-632, 1979.

*6142 - PARMASTO, É.: Rasprostranenie afilloforovykh gribov bazidiosporami 2. Biolo-
giya rasprostraneniya Polyporus squamosus (Polyporaceae). [Dispersion of
aphyllophoraceous fungi by basidiospores 2. Dispersal biology of Polyporus
squamosus (Polyporaceae).] - Izv. Akad. Nauk Eston. SSR, Biol. 27: 139-149,
1978. [In R, ab: Eston, E.]

6143 - PARSONS, J.E., PHENE, C.J., BAKER, D.N., LAMBERT, J.R., McKINION, J.M.: Soil
water stress and photosynthesis in cotton. - Physiol. Plant. 47: 185-189, 1979.

6144 - PARSONS, L.R.: Breeding for drought resistance: What plant characteristics
impart resistance? - HortScience 14: 590-593, 1979.

*6145 - PARTON, W.J., SINGH, J.S., COLEMAN, D.C.: A model of production and turnover
of roots in shortgrass prairie. - J. appl. Ecol. 15: 515-542, 1978.

*6146 - PASICHNYĬ, A.P.: Analiz vodouderzhivayushchikh svoĭstv tkaneĭ rasteniĭ termo-
gravimetricheskim metodom. [Analysis of water-retaining capacity in plant
tissues by the thermogravimetric method.] - Fiziol. Biokhim. kul't. Rast. 10:
318-323, 1978. [In R, ab: E.]

*6147 - PATE, F.M., SNYDER, G.H.: Effect of high water table in organic soil on
yield and quality of forage grasses - lysimeter study. - Soil Crop Sci. Soc.
Florida Proc. 38: 12-14, 1978.

6148 - PATEL, C.L., GHILDYAL, B.P., TOMAR, V.S.: Lysimetery studies on the water use
and growth of rice under different soil-water regimes. - Ind. J. agr. Sci. 49:
90-95, 1979.

*6149 - PATEL, J.D.: How should we interpret and distinguish subsidiary cells? - Bot.
J. Linn. Soc. 77: 65-72, 1978.

*6150 - PATIL, T.T., KHUSPE, V.S.: A study of association between yield and yield
components in Mexican wheat (Triticum aestivum Linn.) variety HD(M) 1553. -
J. Maharashtra agr. Univ. 3: 217-218, 1978.

*6151 - PATIL, T.T., KHUSPE, V.S.: Water use efficiency in Mexican wheat (Triticum
aestivum Linn.) variety HD(M) 1553 (Sonalika) as influenced by nitrogen fertilization and irrigation management. - J. Maharashtra agr. Univ. 3: 219-223,
1978.

*6152 - PATIL, T.T., KHUSPE, V.S.: Influence of nitrogen and irrigation levels on
grain and straw yields, their quality and removal of nitrogen by wheat variety
HD(M) 1553. - Ind. J. Agron. 23: 301-306, 1978.

6153 - PATTERSON, D.T.: The effects of shading on the growth and photosynthetic capacity of itchgrass (Rottboellia exaltata). - Weed Sci. 27: 549-553, 1979.

6154 - PATTERSON, D.T., DUKE, S.O.: Effect of growth irradiance on the maximum photosynthetic capacity of water hyacinth (Eichhornia crassipes (Mart.) Solms). -
Plant Cell Physiol. 20: 177-184, 1979.

6155 - PATTERSON, D.T., FLINT, E.P.: Effects of chilling on cotton (*Gossypium hirsu-tum*), velvetleaf (*Abutilon theophrasti*), and spurred anoda (*Anoda cristata*). - Weed Sci. 27: 473-479, 1979.

*6156 - PAUCĂ-COMĂNESCU, M., TĂCINĂ, A.: Transpiration and water content in some flood plain species. - Rev. Roum Biol., Sér. Biol. vég. 23: 133-144, 1978.

6157 - PAYNE, W.W.: Stomatal patterns in embryophytes: Their evolution, ontogeny and interpretation. - Taxon 28: 117-132, 1979.

*6158 - PEAKE, D.C.I., HENZELL, E.F., STIRK, G.B.: Simulation of herbage production and soil water use by Biloela buffel grass in small plot experiments at Narayen. - Trop. Agron. tech. Memorandum 12: 1-26, 1978.

6159 - PEAKE, D.C.I., HENZELL, E.F., STIRK, G.B., PEAKE, A.: Simulation of changes in herbage biomass and drought response of a buffel grass (*Cenchrus ciliaris* cv. Biloela) in southern Queensland. - Agro-Ecosystems 5: 23-40, 1979.

6160 - PEARSON, C.J.: Daily cycles of photosynthesis, respiration and translocation. - In: MARCELLE, R., CLIJSTERS, H., Van POUCKE, M. (ed.): Photosynthesis and Plant Development. Pp. 125-136. Dr. W. Junk bv Publishers, The Hague - Boston London 1979.

6161 - PEARSON, L.C.: Effects of temperature and moisture on phenology and productiv-ity of Indian ricegrass. - J. Range Manage 32: 127-134, 1979.

*6162 - PECK, A.J.: Development and reclamation of secondary salinity. - In: RUSSELL, J.S., GREACEN, E.L. (ed.): Soil Factors in Crop Production in a Semi-Arid En-vironment. Pp. 301-319. University of Queensland Press, St. Lucia 1977.

6163 - PEMADASA, M.A.: Movements of abaxial and adaxial stomata. - New Phytol. 82: 69-80, 1979.

6164 - PEMADASA, M.A.: Stomatal responses to two herbicidal auxins. - J. exp. Bot. 30: 267-274, 1979.

6165 - PEMADASA, M.A.: Stomatal responses to chemicals. - In: SEN, D.N., CHAWAN, D.D., BANSAL, R.P. (ed.): Structure, Function and Ecology of Stomata. Pp. 143-174. Bishen Singh Mahendra Pal Singh, Dehra Dun 1979.

*6166 - PEMADASA, M.A., JEYASEELAN, K.: Further observations on stomatal responses to sodium azide. - New Phytol. 78: 83-89, 1977.

6167 - PENKA, M., ČERMÁK, J., ŠTĚPÁNEK, V., PALÁT, M.: Diurnal courses of transpira-tion rate and transpiration flow rate as determined by the gravimetric and thermometric methods in a full-grown oak tree (*Quercus robur* L.) - Acta Univ. agr. (Brno), Ser. C 48: 3-30, 1979.

6168 - PENNING de VRIES, F.W.T., WITLAGE, J.M., KREMER, D.: Rates of respiration and of increase in structural dry matter in young wheat, ryegrass and maize plants in relation to temperature, to water stress and to their sugar content. - Ann. Bot. 44: 595-609, 1979.

6169 - PEOPLES, T.R., KOCH, D.W.: Role of potassium in carbon dioxide assimilation in *Medicago sativa* L. - Plant Physiol. 63: 878-881, 1979.

6170 - PEPE, J.F., WELSH, J.R.: Soil water depletion patterns under dryland field conditions of closely related height lines of winter wheat. - Crop Sci. 19: 677-680, 1979.

*6171 - PÉREZ MELIÁN, G., LUQUE, A., CARPENA, O.: Absorción de agua e iones en el cul-tivo de pepinos. II. Relaciones entre los macronutrientes catiónicos. [Water and ions absorption by cucumber plants in hydroponics. II. Cationic macronu-trients relationship.] - Rev. Agroquim. Tecnol Aliment. 18: 245-251, 1978. [In Span, ab: E.]

*6172 - PERRELLE, A.: Modifications des composés pectiques d'*Aster tripolium* sous l'action du sel. - Bull. Soc. bot. France 125 (Actualiés bot. 3/4): 223-228, 1978.

*6173 - PERRIER, A.: Importance des définitions de l'évapotranspiration dans le domaine pratique de la mesure, de l'estimation et de la notion de coefficients cultu-raux. - In: L'Hydrotechnique au Service d'une Politique de l'Eau. Évolution des Problèmes de l'Eau au Cours de la Dernière Décennie. IV (1). Pp. 1-7. Société Hydrotechnique de France, Toulouse 1978.

*6174 - PERRIER, A.: L'évapotranspiration et la photosynthèse d'une culture en fonction de ses propriétés physiques.- Zeszyty problem. Postepów Nauk roln. 203: 485-498, 1978.

6175 - PERRIER, A., HALLAIRE, M.: Rapport de l'évapotranspiration potentielle calculée à l'évaporation mesurée sur bac. I. - Justification d'une relation expérimenta-le obtenue en zone tropicale. - Ann. agron. 30: 329-336, 1979.

6176 - PERRIER, A., HALLAIRE, M.: Rapport de l'évapotranspiration potentielle calculée à l'évaporation mesurée sur bac. II. - Expression en fonction d'un facteur de déséquilibre hydrique entre les surfaces évaporantes et l'air. - Ann. agron. 30: 337-346, 1979.

6177 - PERROUX, K.M.: Controlled water potential in subirrigated pots. - Plant Soil 52: 385-392, 1979.

6178 - PESCHKE, H., MARKGRAF, G.: Über die Wirkung von ^{15}N-markiertem Ammoniumnitrat und Beregnung auf Ertrag, Rohproteingehalt und Aminosäurensusammensetzung von Hafer. - Arch. Acker- Pflanzenbau Bodenk. 23: 547-554, 1979.

6179 - PETERSCHMIDT, N.A., DELANEY, R.H., GREENE, M.C.: Effects of overirrigation on growth and quality of alfalfa. - Agron. J. 71: 752-754, 1979.

*6180 - PETERSON, G.W., WALLA, J.A.: Development of *Dothistroma pini* upon and within needles of Austrian and ponderosa pines in eastern Nebrasca. - Phytopathology 68: 1422-1430, 1978.

*6181 - PETTY, J.A.: Fluid flow through the vessels of birch wood. - J. exp. Bot. 29: 1463-1469, 1978.

*6182 - PFADENHAUER, J.: Contribução ao conhecimento da vegetacão e de suas condicões de crescimento nas dunas costeiras do Rio Grande do Sul, Brasil. [Contribution to the knowledge of the vegetation and its growth conditions on coastal dunes of Rio Grande do Sul, Brazil.] - Rev. Brasil. Biol. 38: 827-836, 1978. [In Port, ab: E.]

6183 - PFADENHAUER, J.: Die Ökologie einiger verbreiteter Dünenpflanzen in Rio Grande do Sul (Südbrasilien) im Hinblick auf ihre Eignung für den Dünenbau. - Bot. Jahrb. Syst. 100: 414-436, 1979.

6184 - PHENE, C.J., FOUSS, J.L., SANDERS, D.C.: Water-nutrient-herbicide management of potatoes with trickle irrigation. - Amer. Potato J. 56: 51-59, 1979.

6185 - PILL, W.G., LAMBETH, V.N., HINCKLEY, T.M.: Effects of Cycocel and nitrogen form on tomato water relations, ion composition, and yield. - Can. J. Plant Sci. 59: 391-397, 1979.

6186 - PINTÉR, L., KÁLMÁN, L., PÁLFI, G.: Determination of drought resistance in maize (*Zea mays* L.) by proline test. - Maydica 24: 155-159, 1979.

6187 - PIRIE, N.W.: The efficiency of protein production by different farming systems. - In: HEWITT, E.J., CUTTING, C.V. (ed.): Nitrogen Assimilation of Plants. Pp. 613-624. Academic Press, London - New York - San Francisco 1979.

*6188 - PITT, M.D., HEADY, H.F.: Responses of annual vegetation to temperature and rainfall patterns in northern California. - Ecology 59: 336-350, 1978.

6189 - PLANCHON, C.: Photosynthesis, transpiration, resistance to CO_2 transfer, and water efficiency of flag leaf of bread wheat, durum wheat and triticale. - Euphytica 28: 403-408, 1979.

6190 - PLANTEFOL, L.: La fourniture d'eau à la plante, dans le cas du Lierre, *Hedera helix* L. - C.R. Acad. Sci. Paris, Sér. D 288: 1549-1554, 1979.

*6191 - PODOLÁK, M.: Vplyv závlahy, hnojenia, hustoty porastu a hybridov na úrodu siláznej kukurice. [Influence of irrigation, fertilizer doses, stand density and hybrids on yield of silage corn.] - Rost. Výroba (Praha) 24: 1163-1171, 1978. [In Slov, ab: R,E,G.]

6192 - POMEROY, M.K., ANDREWS, C.J.: Metabolic and ultrastructural changes associated with flooding at low temperature in winter wheat and barley. - Plant Physiol. 64: 635-639, 1979.

6193 - POWELL, A.A., MATTHEWS, S.: The influence of testa condition on the imbibition and vigour of pea seeds. - J. exp. Bot. 30: 193-197, 1979.

*6194 - PRAKASH, J., BABBER, S., PAHWA, S.K.: Effect of some herbicides on the epidermis of *Vicia faba* (L.) - Weed Res. 18: 379-380, 1978.

6195 - PRASAD, L.K., MUKERJI, S.K.: Herbage growth rate and yield of grasses with supplemental irrigation during dry period in a monsoon. - Ind. J. Agron. 24: 100-101, 1979.

6196 - PRASAD, S.R., MOTIWALE, M.P., SINGH, A.B.: Effect of plant population and moisture regimes on yield and quality of timely and late-planted sugarcane in northern India. - Ind. J. agr. Sci. 49: 262-265, 1979.

6197 - PŘIBÍKOVÁ, E.: K problému měřitelných příznaků pro regulaci obsahu vody a koncentrace minerálních živin v půdě v porostech obilovin. [Measurable signs for the control of water content and mineral element concentration in the soil under cereal stands.] - Rost. Výroba (Praha) 25: 1041-1048, 1979.[In Czech, ab: R,E.G.]

*6198 - PRIEBE, A. JÄGER, H.-J.: Responses of amino acid metabolizing enzymes from plants differing in salt tolerance to NaCl. - Oecologia 36: 307-315, 1978.

*6199 - PRIEBE, A., JÄGER, H.-J.: Einfluss von NaCl auf Wachstum und Ionengehalt unterschiedlich salztoleranter Pflanzen. - Angew. Bot. 52: 331-341, 1978.

6200 - PRIEUR, P., LOUGUET, P.: Activité malate déshydrogénasique NAD-dépendante comparée des extraits d'épidermes de feuilles de *Pelargonium* x *hortorum* à stomates ouverts et fermés. - C.R. Acad. Sci. Paris, Sér. D 288: 947-950, 1979.

6201 - PRIHAR, S.S., SINGH, R., SINGH. N., SANDHU, K.S.: Effects of mulching previous crops or fallow on dryland maize and wheat. - Exp. Agr. 15: 129-134, 1979.

6202 - PUMPHREY, F.V., RAMIG, R.E., ALLMARAS, R.R.: Field response of peas (*Pisum sativum* L.) to precipitation and excess heat. - J. Amer. Soc. hort. Sci. 104: 548-550, 1979.

6203 - PUNZ, W.: The effect of single and combined pollutants on lichen water content. - Biol. Plant. 21: 472-474, 1979.

6204 - PUROHIT, S.S.: [Reversal of abscisic acid induced stomatal closure by kinetin.] - Vijnana Parishad Anusandhan Patrika 22: 327-331, 1979. [In Hindi, ab: E.]

6205 - PUROHIT, S.S.: Stomatal closure by Trifluralin in *Helianthus annuus* L. - Comp. Physiol. Ecol. 4: 17-18, 1979.

6206 - PUZAKOVA, A.I., BATYGIN, N.F., TAT'YANKO, A.K., KOVSHOVA, N.I.: Izuchenie
sravnitel'nogo deĭstviya khimicheskikh i fizicheskikh faktorov na korrelya-
tsionnye zavisimosti i dinamiku fiziologo-biokhimicheskikh protsessov ozimoĭ
pshenitsy. [Studies of a comparative effect of chemical and physical factors
on correlation dependences and dynamics of winter wheat physiological and biochemical processes.] - Fiziol. Biokhim. kul't. Rast. 11: 306-311, 1979. [In R,
ab: E.]

6207 - QUADIR, A., HARRISON, P.J., DeWREEDE, R.E.: The effects of emergence and sub-
emergence of the photosynthesis and respiration of marine macrophytes. -
Phycologia 18: 83-88, 1979.

*6208 - QUARRIE, S.A.: Can abscisic acid be used as a metabolic indicator of drought
resistance in cereals? - In: Proceedings Joint BCPC and BPGRG Symposium: Oppor-
tunities for Chemical Plant Growth Regulation. Pp. 55-61. Reading 1978.

6209 - QUARRIE, S.A., JONES, H.G.: Genotypic variation in leaf water potential, stomatal conductance and abscisic acid concentration in spring wheat subjected
to artificial drought stress. - Ann. Bot. 44: 323-332, 1979.

6210 - QUINTANILLA REJADO, P.: Potassium requirements of cereals. - In: Potassium
Research - Review and Trends. - International Potash Research Institute, Bern
1979.

*6211 - RADEVA, V.: Vliyanie na povishenata atmosferna vlazhnost v"rkhu transpiratsiya-
ta na lyutsernovite rasteniya. [An enhanced air humidity influence on the
lucerne plant transpiration.] - Rasteniev"dni Nauki 15(4): 27-34, 1978. [In
Bulg, ab: R,E.]

6212 - RADEVA, V.: Voden defitsit pri lyutsernata. I. Vliyanie na aerozolnoto orosya-
vane v"rkhu vodniya defitsit i produktivnostta Ĭ. [Water deficit of alfalfa.
I. Effect of aerosolic besprinkling upon the water deficit and productivity of
alfalfa.] - Rasteniev"dni Nauki 16(4): 26-34, 1979. [In Bulg, ab: R,E.]

6213 - RADEVA, V.: Voden defitsit pri lyutsernata. II. Vliyanie na vodniya defitsit
v"rkhu nyakoi fiziologichni pokazateli. [Water deficit of alfalfa. II. Effect
of water deficit on some alfalfa physiological indices.] - Rasteniev"dni Nauki
16(5): 11-19, 1979. [In Bulg., ab: R,E.]

6214 - RADEVA, V., TOPCHIEVA, A.: Vliyanie na periodichnoto pochveno zasushavane
v"rkhu natrupvaneto na sukho veshchestvo, dobiva na semena i nyakoi fiziologichni pokazateli pri lyutsernata. [Periodical soil drought as affecting dry
matter accumulation, seed yield and some physiological characteristics in
alfalfa.] - Rasteniev"dni Nauki 16(1): 13-25, 1979. [In Bulg., ab: R,E.]

6215 - RADIN, J.W., PARKER, L.L.: Water relations of cotton plants under nitrogen deficiency. I. Dependence upon leaf structure. - Plant Physiol. 64: 495-498,
1979.

6216 - RADIN, J.W., PARKER, L.L.: Water relations of cotton plants under nitrogen deficiency. II. Environmental interactions on stomata. - Plant Physiol. 64: 499-
501, 1979.

*6217 - RAGHAVENDRA, A.S., RAO, I.M., DAS, V.S.R.: Replacibility of potassium by sodium
for stomatal opening in epidermal strips of *Commelina benghalensis*. - Z.
Pflanzenphysiol. 80: 36-42, 1976.

6218 - RAMA DAS, V.S., RAJA REDDY, K., KRISHNA, C.M., SAMBA MURTHY, S., RAO, J.V.S.:
Transpirational rates in relation to quality of leaf epicuticular waxes. -
Ind. J. exp. Biol. 17: 158-163, 1979.

6219 - RAMA KRISHNAYYA, G., MURTY, K.S.: Amelioration of drought injury in rice by
chemical sprays. - Curr. Sci. 20: 264-265, 1979.

*6220 - RAMA MOHAN RAO, M.S., RANGA RAO, V., RAMACHANDRAM, M., AGNIHOTRI, R.C.: Effect of vertical mulch on moisture conservation and yield of sorghum in vertisols. - Agr. Water Manage. 1: 333-342, 1978.

*6221 - RAMA MOHAN RAO, M.S., SESHACHALAM, N.: Improvement of intake rates in problem black soils. - Mysore J. agr. Sci. 10: 52-58, 1976.

6222 - RAMATI, A., LIPHSCHITZ, N., WAISEL, Y.: Osmotic adaptation in *Panicum repens*. Differences between organ, cellular and subcellular levels. - Physiol. Plant. 45: 325-331, 1979.

6223 - RAMOS, C., KAUFMANN, M.R.: Hydraulic resistance of rough lemon roots. - Physiol. Plant. 45: 311-314, 1979.

*6224 - RAO, D.V.M., MAHALAKSHMI, B.K., ALI, S.M.: Studies on the relative contribution of various photosynthetic plant parts to the grain development at various moisture levels in *Pennisetum typhoides* (Burm.) S. & H. - Mysore J. agr. Sci. 12: 363-367, 1978.

6225 - RASCHKE, K.: Movements of stomata. - In: HAUPT, W., FEINLEIB, M.E. (ed.): Physiology of Movements. Pp. 383-441. Springer-Verlag. Berlin - Heidelberg - New York 1979.

6226 - RASCHKE, K.: Processes limiting gas exchange: do they optimise production? - In: JOHNSON, C.B. (ed.): Physiological Processes Limiting Plant Productivity. University of Nottingham, Sutton Bonington 1979.

6227 - RASMUSSEN, V.P., KANEMASU, E.T.: Modeling winter wheat yields as affected by water relations and growth regulants. - In: Fourteenth Conference on Agriculture & Forest Meteorology and Fourth Conference on Biometeorology, April 2 - 6, 1979, Minneapolis (Preprint Volume). Pp. 68-69. American Meteorological Society, Boston 1979.

*6228 - RAUNER, Ju.L.: Deciduous forest. - In: MONTEITH, J.L. (ed.): Vegetation and the Atmosphere. Vol. 2. Case Studies. Pp. 241-264. Academic Press, London - New York - San Francisco 1976.

*6229 - RAWITZ, E., HAZAN, A.: The effect of stabilized, hydrophobic aggregate layer properties on soil water regime and seedling emergence. - Soil Sci. Soc. Amer. J. 42: 787-793, 1978.

6230 - RAWSON, H.M.: Vertical wilting and photosynthesis, transpiration, and water use efficiency of sunflower leaves. - Aust. J. Plant Physiol. 6: 109-120, 1979.

6231 - RAY, T.B., BLACK, C.C.: The C_4 pathway and its regulation. - In: GIBBS, M., LATZKO, E. (ed.): Photosynthesis II. Pp. 77-101. Springer-Verlag, Berlin - Heidelberg - New York 1979.

*6232 - REDDY, A.R., DAS, V.S.R.: Changes in stomatal movements in isolated epidermal strips of a CAM plant, *Notonia grandiflora* DC. - Plant Cell Physiol. 19: 1311-1313, 1978.

6233 - REDDY, P.K.R., SHAH, G.L.: Observations on the structure and ontogeny of stomata and trichomes on developing and mature pericarps of *Cassia occidentalis* L. - Biol. Plant. 21: 321-327, 1979.

6234 - REED, R.H., RUSSELL, G.: Adaptation to salinity stress in populations of *Enteromorpha intestinalis* (L.) Link. - Estuarine coastal mar. Sci. 8: 251-258, 1979.

*6235 - REED, R.H., RUSSELL, G.: Salinity fluctuations and their influence on "bottle brush" morphogenesis in *Enteromorpha intestinalis* (L.) Link. - Brit. Phycol. J. 13: 149-153, 1978.

6236 -REETZ, H.F., Jr., HODGES, H.F., DALE, R.F.: Managed soil moisture system for
studying plant water relations under field conditions. - Agron. J. 71: 861-
865, 1979.

6237 - REICOSKY, D.C., DEATON, D.E.: Soybean water extraction, leaf water potential,
and evapotranspiration during drought. - Agron. J. 71: 45-50, 1979.

6238 - REID, C.P.P., BOWEN, G.D.: Effects of soil moisture on V/A mycorrhiza formation
and root development in *Medicago*. - In: HARLEY, J.L., SCOTT RUSSELL, R. (ed.):
The Soil-Root Interface. Pp. 211-219. Academic Press, London - New York - San
Francisco, 1979.

6239 - REID, C.P.P., BOWEN, G.D.: Effect of water stress on phosphorus uptake by my-
corrhizas of *Pinus radiata*. - New Phytol. 83: 103-107, 1979.

6240 - RENARD, C., FLÉMAL, J., BARAMPAMA, D.: Évaluation de la résistance à la sécheresse chez le théier au Burundi. - Café Cacao Thé 23: 175-182, 1979.

*6241 - REUTHER, W.: Citrus. - In: ALVIM, P. de T., KOZLOWSKI, T.T. (ed.): Ecophysi-
ology of Tropical Crops. Pp. 409-439. Academic Press, New York - San Francisco
- London 1977.

*6242 - RHODES, I., STERN, W.R.: Competition for light. - In: WILSON, J.R. (ed.):
Plant Relations in Pastures. Pp. 175-189. CSIRO, Melbourne 1978.

6243 - RICE, J.S., GLENN, E.M., QUISENBERRY, V.L.: A rapid method for obtaining leaf
impressions in grasses. - Agron. J. 71: 894-896, 1979.

6244 - RICHARDSON, C.H., TRUTER, M.R., WINGFIELD, J.N., TRAVIS, A.J., MANSFIELD, T.A.,
JARVIS, R.G.: The effect of benzo-18-crown-6, a synthetic ionophore, on stomatal opening and its interaction with abscisic acid. - Plant Cell Environ. 2:
325-327, 1979.

6245 - RIOU, C., LAGOUARDE, J.P., CHARTIER, R.: Evaporation du sol nu en zone semi-
-aride et en conditions hivernales. Relations avec l'albédo et la température
de la surface du sol. - Ann. agron. 30: 347-362, 1979.

*6246 - RIPLEY, E.A., SAUGIER, B.: Biophysics of a natural grassland: evaporation. -
J. appl. Ecol. 15: 459-479, 1978.

*6247 - RIPLEY, E.A., SAUGIER, B.: Energy and mass exchange of a native grassland in
Saskatchewan. - In: DeVRIES, D.A., AFGAN, N.H. (ed.): Heat and Mass Transfer
in the Biosphere. I. Transfer Processes in Plant Environment. Pp. 311-325.
Halstead Press, New York, John Wiley, New York 1975.

6248 - RIST, D.L., DAVIS, D.D.: The influence of exposure temperature and relative
humidity on the response of pinto bean foliage to sulfur dioxide. - Phytopathology 69: 231-235, 1979.

6249 - RITCHIE, J.T., MEYER, W.S.: Dynamics of water conductance in sorghum roots. -
In: HARLEY, J.L., SCOTT RUSSELL, R. (ed.): The Soil-Root Interface. Pp. 431-
432. Academic Press, London - New York - San Francisco 1979.

6250 - ROBARDS, A.W., CLARKSON, D.T., SANDERSON, J.: Structure and permeability of
the epidermal/hypodermal layers of the sand sedge (*Carex arenaria*, L.). - Protoplasma 101: 331-347, 1979.

6251 - ROBB, J., BRISSON, J.D., BUSCH, L., LU, B.C.: Ultrastructure of wilt syndrome
caused by *Verticillium dahliae*. VII. Correlated light and transmission electron
microscope identification of vessel coatings and tyloses. - Can. J. Bot. 57:
822-834, 1979.

6252 - ROBB, J., SMITH, A., BRISSON, J.D., BUSCH, L.: Ultrastructure of wilt syndrome
caused by *Verticillium dahliae*. VI. Interpretive problems in the study of
vessel coatings and tyloses. - Can. J. Bot. 57: 795-821, 1979.

*6253 - ROBBINS, C.W., BROCKWAY, C.E.: Irrigation water salt concentration influences
on sediment removal by ponds. - Soil Sci. Soc. Amer. J. 42: 478-481, 1978.

6254 - ROBERTS, S.W.: Properties of internal water exchange in leaves of *Illex opaca*
Ait. and *Cornus florida* L. - J. exp. Bot. 30: 955-963, 1979.

6255 - ROBERTS, S.W., KNOERR, K.R., STRAIN, B.R.: Comparative field water relations
of four co-occurring forest tree species. - Can. J. Bot. 57: 1876-1882, 1979.

6256 - ROBINSON, F.E., CUDNEY, D.W., LEHMAN, W.F.: Nitrate fertilizer timing, irrigation, protein, and yellow berry in durum wheat. - Agron. J. 71: 304-308, 1979.

6257 - ROBLIN, G.: Les mouvements révolutifs des feuilles de *Mimosa pudica* L. - Biol.
Plant. 21: 57-65, 1979.

6258 - ROGERS, C.A., POWELL, R.D., SHARPE, P.J.H.: Relationship of temperature to
stomatal aperture and potassium accumulation in guard cells of *Vicia faba*. -
Plant Physiol. 63: 388-391, 1979.

*6259 - ROJAS GARCIDUEÑAS, M., GAMEZ, H.: Efectos del clormequat en cultivares resis-
tentes y susceptibles a sequía de cereales de primavera. [Effect of chlor-
mequat in cereal cultivars resistant and susceptible to drought.] - Turrialba
28: 307-310, 1978. [In Span, ab: E.]

6260 - RÖMHELD, V., MARSCHNER, H.: Fine regulation of iron uptake by the Fe-efficient
plant *Helianthus annuus*. - In: HARLEY, J.L., SCOTT RUSSELL, R. (ed.): The
Soil-Root Interface. Pp. 405-417. Academic Press, London - New York - San
Francisco 1979.

*6261 - ROOK, D.A.: A note on the physiology of Douglas fir. - In: A Review of Douglas
fir in New Zealand. Symposium No. 15. Pp. 201-203. New Zealand Forest Service,
Forest Research Institute, Rotorua 1978.

6262 - ROSE, C.W., DAYANANDA, P.W.A., NIELSEN, D.R., BIGGAR, J.M.: Long-term solute
dynamics and hydrology in irrigated slowly permeable soils. - Irrig. Sci. 1:
77-87, 1979.

6263 - ROSE, D.A.: Soil water: quantities, units and symbols. - J. Soil Sci. 30: 1-15,
1979.

*6264 - ROSENBERG, N.J.: Possible impacts of climatic change and fluctuations on crop
production. - In: GODBY, E.A., OTTERMAN, J. (ed.): COSPAR: The Contribution of
Space Observations Global Food Information Systems. Vol. 2. Pp. 125-136. Pergamon Press, Oxford - New York 1978.

6265 - ROSS, H.A., HEGARTY, T.W.: Sensitivity of seed germination and seedling radicle
growth to moisture stress in some vegetable crop species. - Ann. Bot. 43: 241-
243, 1979.

*6266 - ROSS, J.: Radiative transfer in plant communities. - In: MONTEITH, J.L. (ed.):
Vegetation and the Atmosphere. Vol. 1. Principles. Pp. 13-55. Academic Press,
London - New York - San Francisco 1975.

6267 - ROSS, S.M., TYREE, M.T.: Mason and Maskell's diffusion analogue reconciled with
a translocation theory. - Ann. Bot. 44: 637-640, 1979.

6268 - ROSS WIGHT, J., BLACK, A.L.: Range fertilization: Plant response and water use.
- J. Range Manage. 32: 345-349, 1979.

6269 - ROTH, D.: Beziehungen zwischen dem Bodenfeuchtegehalt und den Erträgen wichti-
ger landwirtschaftlicher Fruchtarten auf zwei unterschiedlichen Schwarzerde-
standorten. - Arch. Acker- Pflanzenbau Bodenk. 23: 429-439, 1979.

6270 - ROTH, D., KACHEL, K.: Untersuchungen zum Einfluss der Beregnung auf den Rüben-,
Blatt- und Rohzuckerertrag der Zuckerrübe auf zwei unterschiedlichen Schwarz-
erdestandorten im Thüringer Becken. - Arch. Acker- Pflanzenbau Bodenk. 23:
185-197, 1979.

*6271 - ROTH, D., TEICHARDT, R., RICHTER, W.: Ergebnisse über Ertragswirksamkeit und
Wasserausnutzung der Beregnung bei wichtigen landwirtschaftlichen Fruchtarten
auf unterschiedlichen Standorten der DDR. - Tagungsber. Akad. Landwirtsch.-
Wiss. DDR, Berlin 1978(166): 475-482, 1978.

6272 - ROWSE, H.R., BARNES, A.: Weather, rooting depth and water relations of broad
beans - a theoretical analysis. - Agr. Meteorol. 20: 381-391, 1979.

*6273 - ROWSE, H.R., STONE, D.A.: Simulation of the water distribution in soil. I.
Measurement of soil hydraulic properties and the model for an uncropped soil. -
Plant Soil 49: 517-531, 1978.

*6274 - ROWSE, H.R., STONE, D.A., GERWITZ, A.: Simulation of the water distribution in
soil. II. The model for cropped soil and its comparison with experiment. -
Plant Soil 49: 533-550, 1978.

6275 - RUBIN, S.S., BAREMBOĬM, A.P.: Vodnyĭ rezhim yabloni pri razlichnykh sistemakh
soderzhaniya pochvy v sadu v usloviyakh Pridnestrov'ya Moldavii. [Water regime of apple-tree with different systems of soil management in orchard under
conditions of the Moldavian Dniester area.] - Fiziol. Biokhim. kul't. Rast.
11: 63-67, 1979. [In R, ab: E.]

6276 - RUDRAPAL, A.B., BASU, R.N.: Physiology of hydration-dehydration treatment in
the maintenance of seed viability in wheat *Triticum aestivum* L. - Ind. J. exp.
Biol. 17: 768-771, 1979.

6277 - RUNDEL, P.W., BRATT, G.C., LANGE, O.L.: Habitat ecology and physiological response of *Sticta filix* and *Pseudocyphellaria delisei* from Tasmania. - Bryo-
logist 82: 171-180, 1979.

*6278 - RUSSELL, G.E.: Varietal differences in formation of sub-stomatal vesicles by
Puccinia striiformis in winter wheat. - Phytopathol. Z. 88: 1-10, 1977.

*6279 - RUSSO, V.M.: Development of *Pinus* seedlings grown from seed subjected to drying
and wetting cycles. - Forest Sci. 24: 537-541, 1978.

6280 - RYCHNOVSKÁ, M. (ed.): Function of Grasslands in Spring Region - Kameničky Project. Progress Report on MAB Project No. 91. - Botanical Institute, Czechoslovak Academy of Sciences, Brno 1979.

6281 - RYCHNOVSKÁ, M.: Physiological variability of *Nardus stricta* in natural and
managed stands. - In: RYCHNOVSKÁ, M. (ed.): Function of Grasslands in Spring
Region - Kameničky Project. Pp. 129-135. Botanical Institute, Czechoslovak
Academy of Sciences, Brno 1979.

6282 - RYCHNOVSKÁ, M.: Transpiration in natural and man-made grasslands at Kameničky.
- In: RYCHNOVSKÁ, M. (ed.): Function of Grasslands in Spring Region - Kameničky
Project. Pp. 153-160. Botanical Institute, Czechoslovak Academy of Sciences,
Brno 1979.

*6283 - RYHINER, A.H., MATSUDA, M.: Effect of plant density and water supply on wheat
production. - Neth. J. agr. Sci. 26: 200-209, 1978.

*6284 - RYSZKOWSKI, L.: Energy and matter economy of ecosystems. - In: Van DOBBEN, W.H.,
LOWE-McCONNEL, R.H. (ed.): Unifying Concepts in Ecology. Pp. 109-126. Dr. W.
Junk b.v. Publ., The Hague; Pudoc, Wageningen 1975.

6285 - SACHS, T.: Cellular interaction in the development of stomatal patterns in *Vinca major* L. - Ann. Bot. 43: 693-700, 1979.

*6286 - SACHS, T., BENOUAICHE, P.: A control of stomata maturation in *Aeonium*. - Isr. J. Bot. 27: 47-53, 1978.

6287 - SAHRAWAT, K.L.: Iron toxicity to rice in an acid sulfate soil as influenced by water regimes. - Plant Soil 51: 143-144, 1979.

6288 - SAKAI, W.S., THOM, M.: Localization of silicon in specific cell wall layers of the stomatal apparatus of sugar cane by use of energy dispersive X-ray analysis. - Ann. Bot. 44: 245-248, 1979

6289 - SALCHEVA, G., GEORGIEVA, D.P.: Vliyanie na pochvenata vlazhnost i mineralnoto khranene prez esenno-zimniya period v"rkhu studoustoĭchivostta na zimnata pshenitsa, otglezhdana na kaneleno-gorska pochva. [Effect of soil moisture and mineral nutrition during the autumn-winter period on the cold resistance of winter wheat grown on cinnamon-forest soil.] - Fiziol. Rast. (Sofia) 5(2): 45-57, 1979. [In Bulg., ab: R,E.]

6290 - SALLÉ, G.: Le système endophytique du *Viscum album*: anatomie et fonctionnement des suçoirs secondaires. - Can. J. Bot. 57: 435-449, 1979.

6291 - SAMMIS, T.W., GAY, L.W.: Evapotranspiration from an arid zone plant community. - J. Arid Environ. 2: 313-321, 1979.

6292 - SAMMIS, T.W., WEEKS, D.L., HANSON, E.G.: Influence of irrigation methods on salt accumulation in row crops. - Trans. ASAE 22: 791-796, 1979.

6293 - SAMMONS, D.J., PETERS, D.B., HYMOWITZ, T.: Screening soybeans for drought resistance. II. Drought box procedure. - Crop Sci. 19: 719-722, 1979.

6294 - SAMSUDDIN, Z., IMPENS, I.: Photosynthesis and diffusion resistances to carbon dioxide in *Hevea brasiliensis* Muell. Agr. clones. - Oecologia 37: 361-363, 1979.

6295 - SAMSUDDIN, Z., IMPENS, I.: The development of photosynthetic rate with leaf age in *Hevea brasiliensis* Muell. Arg. clonal seedlings. - Photosynthetica 13: 267-270, 1979.

6296 - SAMUILOV, F.D., NIKIFOROVA, V.I., NIKIFOROV, E.A.: Issledovanie vliyaniya paramagnitnykh primeseĭ na spin-reshetochnuyu relaksatsiyu protonov vnutrikletochnoĭ vody. [Study of the effect of paramagnetic admixtures on spin-lattice relaxation of intracellular water protons.] - Biofizika 24: 270-273, 1979. [In R, ab: E.]

6297 - SÁNCHEZ-DÍAZ, M.F., MOONEY, H.A.: Resistance to water transfer in desert shrubs native to Death Valley, California. - Physiol. Plant. 46: 139-146, 1979.

*6298 - SANTARIUS, K.A.: Biochemical basis of frost resistance in higher plants. - Acta Hort. 81 (Winter Hardiness in Woody Perennials): 9-21, 1978.

*6299 - SAROOSHI, R.A., ROBERTS, E.A.: Effect of date of pruning, timing, number and composition of oil emulsion sprays on harvest pruned Sultana. - Amer. J. Enol. Vitic. 29: 233-238, 1978.

*6300 - ŞARPE, N., TORGE, C., GANEA, V.: Rezultate privind desicarea ricinului. [Results regarding castor bean desiccation.] - An. Inst. Cercetări Pentru Cereale Plante tehnice Fundulea 43: 319-322, 1978. [In Roum, ab: R,E.]

*6301 - SASTRY, P.S.N.: Evaporation and water balance in a semi-arid monsoonal climate under advective conditions. - In: SETHI, G.R., CHATRATH, M.S. (ed.): Improving Crop and Animal Productivity. Proceedings of International Symposium. Pp. 315-321. Oxford and IBM Publishing Co., New Delhi 1978.

6302 - SATO, K.: [The growth responses of soybean plant to photoperiod and temperature. III. The effects of photoperiod and temperature on the development and
anatomy of photosynthetic organ.] - Jap. J. Crop Sci. 48: 66-74, 1979. [In
Jap, ab: E.]

6303 - SATTER, R.L.: Leaf movements and tendril curling. - In: HAUPT, W., FEINLEIB,
M.E. (ed.): Physiology of Movements. (Encyclopedia of Plant Physiology. N.S.
Vol. 7.) Pp. 442-484. Springer-Verlag, Berlin - Heidelberg - New York 1979.

*6304 - SCANDALIARIS, J., HEMSY, V., RODRIGUEZ MARQUINA, E., LOZANO MUÑOS, H.L., CAJAL,
J.A.: Ciclo de floracionen en mani (Arachis hypogaea L.) y factores que lo
influencian. [Flowering cycle in peanut (Arachis hypogaea L.) and factors which
influence it.] - Rev. agron. Noroeste Argentino 15: 1-54, 1978. [In Span, ab:
E.]

6305 - SCHÄFER, W., DANNOWSKI, M., WURL, B.: Bestimmung des Temperaturoptimums der
CO_2-Aufnahmerate bei Zuckerrüben. - Arch. Acker- Pflanzenbau Bodenk. 23: 289-
296, 1979.

*6306 - SCHEIDECKER, D.: Réactions au sel d'un glycophyte: l'exemple du Haricot. -
Bull. Soc. bot. France 125 (Actualités bot. 3/4): 137-147, 1978.

6307 - SCHINNINGER-ROTHSCHEDL, R.: Der Einfluss isolierter und kombinierter Schad-
stoffe auf Austrocknungsresistenz und Transpiration bei Festuca rubra L. - Z.
Pflanzenphysiol. 94: 351-362, 1979.

*6308 - SCHNETTER, M.L.: Der Einfluss von Aussenfaktoren auf die Struktur des Blattes
von Avicennia germinans (L.) L. unter natürlichen Bedingungen. - Beitr. Biol.
Pflanzen 54: 13-28, 1978.

6309 - SCHOBERT, B.: Die Akkumulierung von Prolin in Phaeodactylum tricornutum und
die Funktion der "compatible solutes" in Pflanzenzellen unter Wasserstress. -
Ber. Deut. bot. Ges. 92: 23-30, 1979.

6310 - SCHÖNHERR, J., ECKL, K., GRULER, H.: Water permeability of plant cuticles: The
effect of temperature on diffusion of water. - Planta 147: 21-26, 1979.

6311 - SCHÖNHERR, J., SCHMIDT, H.W.: Water permeability of plant cuticles. Dependence
of permeability coefficients of cuticular transpiration on vapor pressure saturation deficit. - Planta 144: 391-400, 1979.

6312 - SCHULZE, E.D., HALL, A.E.: Short-term and long-term effects of drought on transpiration, assimilation, biomass production, and yield. - In: JOHNSON, C.B.
(ed.): Physiological Processes Limiting Plant Productivity. University of
Nottingham, Sutton Bonington 1979.

6313 - SCHULZE, E.-D., KÜPPERS, M.: Short-term and long-term effects of plant water
deficits on stomatal response to humidity in Corylus avellana L. - Planta 146:
319-326, 1979.

6314 - SCHURER, K., GRIFFIOEN, H., KORNET, J.G., VISSCHER, G.J.W.: Measurement of the
rate of water flow in plants. - Neth. J. agr. Sci. 27: 136-141, 1979.

*6315 - SCHWARZ, K., ROTH, D., TEICHARDT, R., BERGER, W.: Produktivität und Effektivi-
tät der Beregnung in Abhängigkeit von Fruchtart, Standort und Bodennutzungstyp.
- Arch. Acker- Pflanzenbau Bodenk. 22: 721-730, 1978.

6316 - SCOTT, H.D., BATCHELOR, J.T.: Dry weight and leaf area production rates of irrigated determinate soybeans. - Agron. J. 71: 776-782, 1979.

6317 - SCOTT, H.D., GEDDES, R.D.: Plant water stress of soybean (Glycine max) and
common cocklebur (Xanthium pensylvanicum): A comparison under field conditons.
- Weed Sci. 27: 285-289, 1979.

6318 - SEFA-DEDEH, S., STANLEY, D.W.: The relationship of microstructure of cowpeas
to water absorption and dehulling properties. - Cereal Chem. 56: 379-386, 1979.

6319 - SEGINER, I,: Irrigation uniformity related to horizontal extent of root zone.
A computation study. - Irrig. Sci. 89-96, 1979.

6320 - SELIRIO, I.S., BROWN, D.M.: Soil moisture-based simulation of forage yield. -
Agr. Meteorol. 20: 99-114, 1979.

6321 - SEN, D.N., CHAWAN, D.D., BANSAL, R.P. (ed.): Structure, Function and Ecology
of Stomata. - Bishen Singh Mahendra Pal Singh, Dehra Dun 1979.

6322 - SEN, D.N., CHAWAN, D.D., BANSAL, R.P.: Stomatal studies on arid zone plants. -
In: SEN, D.N., CHAWAN, D.D., BANSAL, R.P. (ed.): Structure, Function and Ecology of Stomata. Pp. 217-228. Bishen Singh Mahendra Pal Singh, Dehra Dun 1979.

6323 - SEPASKHAH, A.R., BOERSMA, L.: Elongation of wheat leaves exposed to several
levels of matric potential and NaCl-induced osmotic potential of soil water. -
Agron. J. 71: 848-852, 1979.

6324 - SEPASKHAH, A.R., BOERSMA, L.: Shoot and root growth of wheat seedlings exposed
to several levels of matric potential and NaCl-induced osmotic potential of
soil water. - Agron. J. 71: 746-752, 1979.

6325 - SETTER, T.L., GREENWAY, H.: Growth and osmoregulation of *Chlorella emersonii*
in NaCl and neutral osmotica. - Aust. J. Plant Physiol. 6: 47-60, 1979.

6326 - SHACKEL, K.A., HALL, A.E.: Reversible leaflet movements in relation to drought
adaptation of cowpeas, *Vigna unguiculata* (L.) Walp. - Aust. J. Plant Physiol.
6: 265-276, 1979.

*6327 - SHALHEVET, J., BIELORAI, H.: Crop water requirement in relation to climate and
soil. - Soil Sci. 125: 240-247, 1978.

6328 - SHAPOSHNIKOVA, S.V.: Vydelenie khromatina iz prorostkov pshenitsy raznymi meto-
dami i vozmozhnost' izucheniya ego funktsional'nogo sostoyaniya pri obezvozhi-
vanii. [Isolation of chromatin from wheat seedlings by various methods and a
possiblity of investigation its functional state in plants under conditions
of dehydration.] - Fiziol. Rast. 26: 179-183, 1979. [In R, ab: E.]

*6329 - SHARMA, J.P., SHARMA, K.D.: Studies on moisture content, mycoflora and germi-
nation of lentil (*Lens esculenta* Moench.) seeds. - Seed Res. 6: 31-37, 1978.

6330 - SHARMA, N.S., VARGHESE, S., DESAI, J., CHINOY, J.J.: Biosynthesis of solasodine
in developing berries of *Solanum khasianum* Clarke. - Ind. J. exp. Biol. 17:
224-225, 1979.

6331 - SHARMA, R.P., PARASHAR, K.S.: Effect of different water supplies, levels of N
and P on consumptive use of water, water use efficiency and moisture extraction
pattern by cauliflower (var. Snowball-16). - Ind. J. Agron. 24: 315-321, 1979.

6332 - SHARP, R.E., DAVIES, W.J.: Solute regulation and growth by roots and shoots of
water-stressed maize plants. - Planta 147: 43-49, 1979.

6333 - SHARP, R.E., OSONUBI, O., WOOD, W.A., DAVIES, W.J.: A simple instrument for
measuring leaf extension in grasses, and its application in the study of the
effects of water stress on maize and sorghum. - Ann. Bot. 44: 35-45, 1979.

6334 - SHEEHY, J.E., WOODWARD, F.I., JONES, M.B., WINDRAM, A.: Microclimate, photosynthesis and growth of lucerne (*Medicago sativa* L.). I. Microclimate and
photosynthesis. - Ann. Bot. 44: 693-707, 1979.

6335 - SHELDRAKE, A.R.: Effects of osmotic stress on polar auxin transport in *Avena*
mesocotyl sections. - Planta 145: 113-117, 1979.

6336 - SHERIFF, D.W.: Water vapour and heat transfer in leaves. - Ann. Bot. 43: 157-171, 1979.

6337 - SHIMADA, K., OGAWA, T., SHIBATA, K.: Isotachophoretic analysis of ions in guard cells of *Vicia faba*. - Physiol. Plant. 47: 173-176, 1979.

6338 - SHIMSHI, D.: Leaf permeability as an index of water relations, CO_2 uptake and yield of irrigated wheat. - Irrig. Sci. 1: 107-117, 1979.

6339 - SHMAT'KO, I.G., GULYAEV, B.I., SHVEDOVA, O.E., GOLIK, K.N., LATASHENKO, O.P.: Parametry vodnogo rezhima i gazoobmena sortov ozimoĭ pshenitsy pri ukhudshenii vodoobespechennosti. [Parameters of water regime and gas exchange in winter wheat varieties under decreasing water supply.] - Fiziol. Biokhim. kul't. Rast. 11: 312-317, 332, 1979. [In R, ab: E.]

6340 - SHMAT'KO, I.G., SHVEDOVA, O.E.: Regulyatornye faktory vodoobmena kul'turnykh rasfeniĭ. [Regulatory factors of cultivated plant water exchange.] - Fiziol. Biokhim. kul't. Rast. 11: 460-470, 1979. [In R, ab: E.]

6341 - SHOKES, F.M., McCARTER, S.M.: Occurrence, dissemination, and survival of plant pathogens in surface irrigation ponds in Southern Georgia. - Phytopathology 69: 510-516, 1979.

6342 - SHOMER-ILAN, A., NISSENBAUM, A., GALUN, M., WAISEL, Y.: Effect of water regime on carbon isotope composition of lichens. - Plant Physiol. 63: 201-205, 1979.

6343 - SIDDARAMAPPA, R., WATANABE, I.: Evidence for vapor loss of ^{14}C-carbofuran from rice plants. - Bull. Environ. Contam. Toxicol. 23: 544-551, 1979.

6344 - SIEVERDING, E.: Einfluss der Bodenfeuchte auf die Effektivität der VA-Mykorrhiza. - Angew. Bot. 53: 91-98, 1979.

6345 - SIMINOVITCH, D.: Protoplasts surviving freezing to -196 C and osmotic dehydration in 5 molar salt solutions prepared from the bark of winter black locust trees. - Plant Physiol. 63: 722-725, 1979.

6346 - SIMMONS, C.S., NIELSEN, D.R., BIGGAR, J.W.: Scaling of field-measured soil-water properties. I. Methodology. II. Hydraulic conductivity and flux. - Hilgardia 47: 77-173, 1979.

*6347 - ŠIMON, J.: Různá organizace porostu zavlažované cukrovky na lehké půdě. [Various methods of organization of the irrigated sugar beet stand on light soil.] - Rost. Výroba (Praha) 24: 1183-1192, 1978. [In Czech, ab: R,E,G.]

*6348 - SIMONS, J.: Verbreitung und Dynamik von *Vaucheria*-Algengesellschaften in den südwestlichen Niederlanden. - Vegetatio 38: 119-122, 1978.

6349 - SIMPSON, G.M., DURLEY, R.C., KANNANGARA, T., STOUT, D.G.: The problem of plant breeders. - In: SCOTT, T.K. (ed.): Plant Regulation and World Agriculture. Pp. 111-128. Plenum Press, New York - London 1979.

6350 - SIMPSON, J.R., PINKERTON, A., LAZDOVSKIS, J.: Interacting effects of subsoil acidity and water on the root behaviour and shoot growth of some genotypes of lucerne (*Medicago sativa* L.) - Aust. J. agr. Res. 30: 609-619, 1979.

*6351 - SIMS, P.L., SINGH, J.S.: The structure and function of ten western North American grasslands. II. Intra-seasonal dynamics in primary producer compartments. - J. Ecol. 66: 547-572, 1978.

*6352 - SIMS, P.L., SINGH, J.S.: The structure and function of ten western North American grasslands. III. Net primary production, turnover and efficiencies of energy capture and water use. - J. Ecol. 66: 573-597, 1978.

*6353 - SIMS, P.L., SINGH, J.S., LAUENROTH, W.K.: The structure and function of ten western North American grasslands. I. Abiotic and vegetational characteristics. - J. Ecol. 66: 251-285, 1978.

6354 - SIMS, R.E.H.: Comparative methods of harvesting oilseed rape. - N. Zeal. J. exp. Agr. 7: 79-83, 1979.

6355 - SIMS, R.E.H.: Drying cycles and optimum harvest stage of oilseed rape. - N. Zeal. J. exp. Agr. 7: 85-89, 1979.

6356 - SINCLAIR, D.F., WILLIAMS, J.: Components of variance involved in estimating soil water content and water content change using a neutron moisture meter. - Aust. J. Soil.Res. 17: 237-247, 1979.

6357 - SINCLAIR, T.R., JOHNSON, M.N., DRAKE, G.M., Van HOUTTE, R.C.: Mobile laboratory for continuous, long-term gas exchange measurements of 39 leaves. - Photosynthetica 13: 446-453, 1979.

*6358 - SINGH, A.: Root development of crops under irrigation in the arid zone. - Ann. Arid Zone 17: 370-376, 1978.

6359 - SINGH, A.: Moisture fluctuations, moisture use and water balance under principal land use systems of arid region. - Ann. Arid Zone 18: 80-85, 1979.

6360 - SINGH, B.N., GALSTON, E., DASHEK, W., WALTON, D.C.: Abscisic acid levels and metabolism in the leaf epidermal tissue of *Tulipa gesneriana* L. and *Commelina communis* L. - Planta 146: 135-138, 1979.

*6361 - SINGH, J., MILLER, R.W.: Spin label studies of membranes in winter rye protoplasts during freezing. - Plant Physiol. 61(Suppl.): 5, 1978.

6362 - SINGH, L.B.: Mango. - In: ALVIM, P. de T., KOZLOWSKI, T.T. (ed.): Ecophysiology of Tropical Crops. Pp. 479-485. Academic Press, New York - San Francisco - London 1977.

6363 - SINGH, N.T., SINGH, R., MAHAJAN, P.S., VIG, A.C.: Influence of supplemental irrigation and pre-sowing soil water storage on wheat. - Agron. J. 71: 483-489, 1979.

6364 - SINGH, N.T., VIG, A.C., SINGH, R., CHAUDHARY, M.R.: Influence of different levels of irrigation and nitrogen on yield and nutrient uptake by wheat. - Agron. J. 71: 401-404, 1979.

6365 - SINGH, P., RUSSELL, M.B.: Water balance and profile moisture loss patterns of an alfisol. - Agron. J. 71: 963-966, 1979.

6366 - SINGH, R., DASS, B., SINGH, N., SINGH, Y.: Effect of N fertilization on yield and moisture extraction by rainfed maize as affected by soil type and rainfall in Punjab, India. - Field Crops Res. 2: 109-115, 1979.

*6367 - SINGH, R., VERMA, H.N., SINGH, P., SINGH, N.: Performance of different varieties of gram, barley and rape and mustard in rainfed sub-humid conditions of Punjab. - Ind. J. Agron. 23: 355-358, 1978.

*6368 - SINGH, R.K., DE, R.: Dry-matter accumulation and nitrogen uptake in wheat as affected by nitrogen application at varying moisture regimes. - J. Ind. Soc. Soil Sci. 26: 363-366, 1978.

6369 - SINGH, S.D., YUSUF, M.: Effect of water, nitrogen and row spacing on yield and oil content of brown sarson. - Can. J. Plant Sci. 59: 437-444, 1979.

6370 - SINHA, R.N., WHITE, N.D.G., WALLACE, H.A.H., McKENZIE, R.I.H.: Effect of moisture content on viability and infestation of hulless Terra oats in storage. - Can. J. Plant Sci. 59: 911-916, 1979.

6371 - SIRYK, A.A.: Vikova dynamika lystkovoi masy derevnykh porid pry zroshenni.
[Age dynamics of woody plants leaf mass under irrigation.] - Ukr. bot. Zh. 36:
267-268, 1979. [In Ukr., ab: E,R.]

6372 - SITHAMPARANATHAN, J.: Improvement of winter forage production on high rainfall
North Island hill country by autumn oversowing of cereals. I. Relative performance of different winter cereals cultivars. - N. Zeal.J. exp. Agr. 7: 281-
284, 1979.

6373 - SITHAMPARANATHAN, J.: Improvement of winter forage production on high rainfall
North Island hill country by autumn oversowing of cereals. II. Effects of
oversowing date and nitrogen application. - N. Zeal. J. exp. Agr. 7: 285-288,
1979.

6374 - SIVAKUMAR, M.V.K., SEETHARAMA, N., SINGH, S., BIDINGER, F.R.: Water relations,
growth, and dry matter accumulation of sorghum under post-rainy season conditions. - Agron. J. 71: 843-847, 1979.

6375 - SIVAKUMAR, M.V.K., SHAI!, R.H.: Attenuation of radiation in moisture-stressed
and unstressed soybeans. - Iowa State J. Res. 53: 251-257, 1979.

6376 - SIVAKUMAR, M.V.K., SHAW, R.H.: Stomatal conductance and leaf-water potential
of soybeans under moisture stress. - Iowa State J. Res. 54: 17-27, 1979.

6377 - ŠKODA, V.: Regulace bioenergetického potenciálu půdy minerálními hnojivy a
předpoklady jejich využití v závlahových podmínkách u cukrové řepy. [Bio-ener-
getic potential of land controlled by mineral fertilizers and the prerequisites
of their application to irrigated sugar-beet.] - Rost. Výroba (Praha) 25: 1031
-1039, 1979. [In Czech, ab: R,E,G.]

6378 - SLABBERS, P.J., SORBELLO HERRENDORF, V., STAPPER, M.: Evaluation of simplified
water-crop yield models. - Agr. Water Manage. 2: 95-129, 1979.

6379 - SLAVÍK, L.: Regulační funkce doplňkových závlah při stabilizaci výnosu ozimé
pšenice. [Regulatory function of supplemental irrigation in the stabilization
of winter wheat yields.] - Rost. Výroba (Praha) 25: 1023-1030, 1979. [In Czech,
ab: R,E,G.]

6380 - SLOWIK, K., LABANAUSKAS, C.K., STOLZY, L.H., ZENTMYER, G.A.: Influence of rootstocks, soil oxygen, and soil moisture on the uptake and translocation of nutrients in young avocado plants. - J. Amer. Soc. hort. Sci. 104: 172-175, 1979.

6381 - SLUKHAĬ, S.I., KIRICHENKO, V.P.: Vodnye resursy i vodopotreblenie polevykh
kul'tur. [Water resources and water intake of field crops.] - Fiziol. Biokhim.
kul't. Rast. 11: 471-481, 1979. [In R, ab: E.]

6382 - SLUKHAĬ, S.I., KIRICHENKO, V.P., PETRENKO, N.I.: Vodopotreblenie i produktivnost' kukuruzy pri optimizatsii nekotorykh faktorov rosta. [Water consumption
and productivity of maize under optimization some of growth factors.] - Fiziol.
Biokhim. kul't. Rast. 11: 324-332, 1979. [In R, ab: E.]

6383 - ŠMÍD, P.: Evapotranspiration from a natural grassland stand, estimated by the
heat-balance method. - In: RYCHNOVSKÁ, M. (ed.): Function of Grasslands in
Spring Region - Kameničky Project. Pp. 51-57. Botanical Institute, Czechoslovak
Academy of Sciences, Brno 1979.

6384 - SMITH, C.W., WADDLE, B.A., RAMEY, H.H., Jr.: Plant spacings with irrigated cotton. - Agron. J. 71: 858-860, 1979.

6385 - SMITH, M.W., KENWORTHY, A.L.: The response of fruit trees in Michigan to
trickle irrigation. - Commun. Soil Sci. Plant Anal. 10: 1371-1380, 1979.

6386 - SMITH, W.K., GELLER, G.N.: Plant transpiration at high elevations: Theory,
field measurements, and comparisons with desert plants. - Oecologia 41: 109-
122, 1979.

6387 - SO, H.B.: Water potential gradients and resistances of a soil-root system measured with the root and soil psychrometer. - In: HARLEY, J.L., SCOTT RUSSELL, R. (ed.): The Soil-Root Interface. Pp. 99-113. Academic Press, London - New York - San Francisco 1979.

6388 - SO, H.B.: An analysis of the relationship between stem diameter and leaf water potentials. - Agron. J. 71: 675-679, 1979.

6389 - SO, H.B., REICOSKY, D.C., TAYLOR, H.M.: Utility of stem diameter changes as predictors of plant canopy water potential. - Agron. J. 71: 707-713, 1979.

*6390 - SOBERALSKE, R.M., ANDREW, R.H.: Gene effects on kernel moisture and sugars of near-isogenic lines of sweet corn. - Crop Sci. 18: 743-746, 1978.

6391 - SOJKA, R.E., STOLZY, L.H., FISCHER, R.A.: Comparison of diurnal drought response of selected wheat cultivars. - Agron. J. 71: 329-335, 1979.

6392 - SOLÁROVÁ, J., POSPÍŠILOVÁ, J.: Diffusive conductances of adaxial and abaxial epidermes: Response to photon flux density during development of water stress in primary bean leaves. - Biol. Plant. 21: 446-451, 1979.

6393 - SOLDATINI, G.F.: Effetti metabolici indotti da stress idrico in piantine di mais trattate con polietilenglicol. [Metabolic effects induced by water stress in corn crops treated with polyethylene glycol.] - Agrochimica 23: 367-376, 1979. [In Ital, ab: E,G,F,Span.]

6394 - SOLIDAY, C.L., KOLATTUKUDY, P.E., DAVIS, R.W.: Chemical and ultrastructural evidence that waxes associated with the suberin polymer constitute the major diffusion barrier to water vapor in potato tuber (Solanum tuberosum L.). - Planta 146: 607-614, 1979.

6395 - SOMERS, G.F.: Production of food plants in areas supplied with highly saline water: Problems and prospects. - In: MUSSELL, H., STAPLES, R. (ed.): Stress Physiology in Crop Plants. Pp. 107-125. John Wiley & Sons, Inc., New York 1979.

6396 - SOMERS, G.F., FONTES, M., GRANT, D.M.: Halophytes from coastal salt marshes: A potential source of crop plants for arid lands. - In: GOODIN, J.R., NORTHINGTON, D.K. (ed.): Arid Land Plant Resources. Pp. 402-417. Texas Technical University, Lubbock 1979.

6397 - SOMERVILLE, S.C., BRIGGS, K.G.: Evaluation of maturity assessments in wheat and barley. - Can. J. Plant Sci. 59: 505-510, 1979.

*6398 - SOMMER, C., BRAMM, A.: Wasserverbrauch und Pflanzenwachstum bei Zuckerrüben in Abhängigkeit von der Wasserversorgung. - Landbauforsch. Völkenrode 28: 151-158, 1978.

6399 - SPALDING, M.H., STUMPF, D.K., KU, M.S.B., BURRIS, R.H., EDWARDS, G.E.: Crassulacean acid metabolism and diurnal variations of internal CO_2 and O_2 concentrations in Sedum praealtum DC. - Aust. J. Plant Physiol. 6: 557-567, 1979.

6400 - SPOTTS, R.A., FERREE, D.C.: Effect of overtree misting for bloom delay on Venturia inaequalis ascospore maturity and release. - Plant Dis. Rep. 63: 108-112, 1979.

6401 - SPOTTS, R.A., FERREE, D.C.: Photosynthesis, transpiration, and water potential of apple leaves infected by Venturia inaequalis. - Phytopathology 69: 717-719, 1979.

6402 - SPYROPOULOS, C.G., LAMBIRIS, M.P.: Influence of temperature on the effects of water stress on Quercus species. - Ann. Bot. 44: 215-220, 1979.

6403 - SQUIRE, G.R.: The response of stomata of pearl millet (*Pennisetum typhoides* S. and H.) to atmospheric humidity. - J. exp. Bot. 30: 925-933, 1979.

6404 - STAHLMAN. P.W., PHILLIPS, W.M.: Effects of water quality and spray volume on glyphosate phytotoxicity. - Weed Sci. 27: 38-41, 1979.

6405 - STALL, R.E., COOK, A.A.: Evidence that bacterial contact with the plant cell is necessary for the hypersensitive reaction but not the susceptible reaction. - Physiol. Plant Pathol. 14: 77-84, 1979.

6406 - STECKEL, J.R.A., GRAY, D.: Drought tolerance in potatoes. - J. agr. Sci. 92: 375-381, 1979.

*6407 - STEIN, J., PLAMONDON, A.P.: Calibration de l'atmomètre Bellani par la méthode de Penman, Luceville, Québec. - Naturaliste Can. 105: 467-471, 1978.

6408 - STEINECK, O., HAEDER, H.E.: The effect of potassium on growth and yield components of plants. - In: Potassium Research - Review and Trends. Pp. 165-187. International Potash Research Institute, Bern 1979.

6409 - STEPONKUS, P.L.: Effects of freezing and cold acclimation on membrane structure and function. - In: MUSSELL, H., STAPLES, R. (ed.): Stress Physiology in Crop Plants. Pp. 144-158. John Wiley & Sons, Inc., New York 1979.

6410 - STEVENS, R.A., MARTIN, E.S.: The structure of guard cells and substomatal ion-adsorbent bodies. - In: SEN, D.N., CHAWAN, D.D., BANSAL, R.P. (ed.): Structure, Function and Ecology of Stomata. Pp. 7-21. Bishen Singh Mahendra Pal Singh, Dehra Dun 1979.

*6411 - STILLE, B., UZELAC, G.: Das Verhalten von Mikroorganismen in Abhängigkeit von der Wasseraktivität. - Z. Pflanzenkrankheiten Pflanzenschutz 85: 186-190, 1978.

6412 - STOCK, H.-G., DÖRFEL, H., ZIEGLER, G.: Untersuchungen zur Genauigkeit von gravimetrischen Bodenfeuchtemessungen. - Arch. Acker- Pflanzenbau Bodenk. 23: 399-407, 1979.

*6413 - STONE, J.F.: Evapotranspiration control on agricultural lands. - Soil Crop Sci. Soc. Florida Proc. 37: 1-11, 1978.

*6414 - STONE, L.R., GWIN, R.E., Jr., DILLON, M.A.: Corn and grain sorghum yield response to limited irrigation. - J. Soil Water Conserv. 33: 235-238, 1978.

*6415 - STONER, W.A., MILLER, P.C., OECHEL, W.C.: Simulation of the effect of the tundra vascular plant canopy on the productivity of four moss species. - In: TIESZEN, L.L.(ed.): Vegetation and Production Ecology of an Alaskan Arctic Tundra. Pp. 371-387. Springer-Verlag, New York - Heidelberg - Berlin 1978.

6416 - STOREY, R., WYN JONES, R.G.: Responses of *Atriplex spongiosa* and *Suaeda monoica* to salinity. - Plant Physiol. 63: 156-162, 1979.

*6417 - STOUT, D.G., STEPONKUS, P.L., COTTS, R.M.: Plasmalemma alteration during cold acclimation of *Hedera helix* bark. - Can. J. Bot. 56: 196-205, 1978.

*6418 - STURGES, D.L., TRLICA, M.J.: Root weights and carbohydrate reserves of big sagebrush. - Ecology 59: 1282-1285, 1978.

6419 - STUTTE, C.A., WEILAND, R.T., BLEM, A.R.: Gaseous nitrogen loss from soybean foliage. - Agron. J. 71: 95-97, 1979.

*6420 - SUDNITSYNA, T.N.: Vliyanie nedostatochnoǐ vlagoobespechenosti na azotnoe pitanie sosnovykh kul'tur. [Effect of insufficient water-supply on nitrogen nutrition of pine cultures.] - Lesovedenie 1978 (4): 60-68, 1978. [In R, ab: E.]

6421 - SUGIYAMA, T., SCHMITT, M.R., KU, S.B., EDWARDS, G.E.: Differences in cold ' lability of pyruvate, Pi dikinase among C₄ species. - Plant Cell Physiol. 20: 965-971, 1979.

6422 - SULLIVAN, C.Y., ROSS, W.M.: Selecting for drought and heat resistance in grain sorghum. - In: MUSSELL, H., STAPLES, R. (ed.): Stress Physiology in Crop Plants. Pp. 264-281. John Wiley & Sons, Inc., New York 1979.

6423 - SUMAYAO, C.R., KANEMASU, E.T.: Temperature and stomatal resistance of soybean leaves. - Can. J. Plant Sci. 59: 153-162, 1979.

6424 - SUNDQVIST, C., RYBERG, H.: Structure of protochlorophyll-containing plastids in the inner seed coat of pumpkin seeds (*Cucurbita pepo*). - Physiol. Plant. 47: 124-128, 1979.

6425 - SUNG, F.J.M., KRIEG, D.R.: Relative sensitivity of photosynthetic assimilation and translocation of ^{14}carbon to water stress. - Plant Physiol. 64: 852-856, 1979.

6426 - ŠVACHULA, V., ŠVACHULOVÁ, J.: Vliv extrémních vláhových podmínek na změny obsahu chlorofylů v listech cukrovky. [The effect of extreme moisture conditions on changes in the content of chlorophyll in sugar beet leaves.] - Rost. Výroba (Praha) 25: 1-8, 1979. [In Czech, ab: E,G,R.]

*6427 - SVESHNIKOVA, V.M.: O vodnom rezhime *Haloxylon ammodendron* (C.A.Mey.) v Severnoĭ Gobi i Karakumakh. [About *Haloxylon ammodendron* (C.A.Mey) water regime in the Northern Gobi and Karakum deserts.] - Problemy Osvoeniya Pustyn' 1978 (1): 77-79, 1978. [In R, ab: E.]

6428 - SVESHNIKOVA, V.M.: Dominanty Kazakhstanskikh Stepeĭ. Ėkologo-fiziologicheskaya Kharakteristika. [Dominants of Steppes in Kazakhstan. Ekological and Physiological Characteristics.] - Nauka, Leningrad 1979. [In R.]

6429 - SWEENEY, B.M.: Endogenous rhythms in the movement of plants. - In: HAUPT, W., FEINLEIB, M.E. (ed.): Physiology of Movements. Pp. 71-93. Springer-Verlag, Berlin - Heidelberg - New York 1979.

6430 - SYVERTSEN, J.P., CUNNINGHAM, G.L.: The effects of irradiating adaxial or abaxial leaf surface on the rate of net photosynthesis of *Perezia nana* and *Helianthus annuus*. - Photosynthetica 13: 287-293, 1979.

6431 - SZUJKÓ-LACZA, J., SEN, S., HORVÁTH, I.: Effect of different light intensities on the anatomical characteristics of the leaves of *Pimpinella anisum* L. - Acta agron. Acad. Sci. Hung. 28: 120-131, 1979.

*6432 - TAI, E.A.: Banana. - In: ALVIM, P. de T., KOZLOWSKI, T.T. (ed.): Ecophysiology of Tropical Crops. Pp. 441-460. Academic Press, New York - San Francisco - London 1977.

6433 - TAKAHAMA, U.: Stimulation of lipid peroxidation and carotenoid bleaching by deuterium oxide in illuminated chloroplast fragments: Participation of singlet molecular oxygen in the reactions. - Plant Cell Physiol. 20: 213-218, 1979.

6434 - TAL, M., IMBER, D., EREZ, A., EPSTEIN, E.: Abnormal stomatal behavior and hormonal imbalance in *flacca*, a wilty mutant of tomato. V. Effect of abscisic acid on indoleacetic acid metabolism and ethylene evolution. - Plant Physiol. 63: 1044-1048, 1979.

6435 - TAL, M., KATZ, A., HEIKIN, H., DEHAN, K.: Salt tolerance in the wild relatives of the cultivated tomato: Proline accumulation in *Lycopersicon esculentum* Mill., *L. peruvianum* Mill. and *Solanum pennelli* Cor. treated with NaCl and polyethylene glycole. - New Phytol. 82: 349-355, 1979.

6436 - TAL, M., ROSENTAL, I., ABRAMOVITZ, R., FORTI, M.: Salt tolerance in *Simmondsia chinensis*: Water balance and accumulation of chloride, sodium and proline under low and high salinity. - Ann. Bot. 43: 701-708, 1979.

6437 - TASHAKORIE, A., PHAM THI, A.T., VIEIRA da SILVA, J.: La glycolate oxydase chez le Soja (*Glycine max*. L. Merr.). Quelques caractéristiques et influence du déficit hydrique. - C.R. Acad. Sci. Paris, Sér. D 289: 1101-1104, 1979.

6438 - TAYLOR, G.B., PALMER, M.J.: The effect of some environmental conditions on seed development and hard-seededness in subterranean clover (*Trifolium subterraneum* L.) - Aust. J. agr. Res. 30: 65-76, 1979.

*6439 - TAZAKI, T., USHIJIMA, T.: General discussion of photosynthesis of single leaves. - In: MONSI, M., SAEKI, T., (ed.): Ecophysiology of Photosynthetic Productivity. JIBP Synthesis, Vol. 19. Pp. 72-75. University Tokyo Press, Tokyo 1978.

6440 - TEKINEL, O.: Water stress and its implications (irrigation) in the future of agriculture. - In: SCOTT, T.K. (ed): Plant Regulation and World Agriculture. Pp. 457-473. Plenum Press, New York - London 1979.

6441 - TERAMURA, A.H., DAVIES, F.S., BUCHANAN, D.W.: Comparative photosynthesis and transpiration in excised shoots of rabbiteye blueberry. - HortScience 14: 723-724, 1979.

6442 - TERAMURA, A.H., STRAIN, B.R.: Localized populational differences in the photosynthetic response to temperature and irradiance in *Plantago lanceolata*. - Can. J. Bot. 57: 2559-2563, 1979.

*6443 - TETLOW, R.M., FENLON, J.S.: Pre-harvest desiccation of crops for conservation. 1. Effect of steam and formic acid on the moisture concentration of lucerne, ryegrass and tall fescue before and after cutting. - J. Brit. Grassl. Soc. 33: 213-222, 1978.

6444 - THIAGARAJAH, M.R., HUNT, L.A., HUNTER, R.B.: Effects of short-term temperature fluctuations on leaf photosynthesis in corn (*Zea mays*). - Can. J. Bot. 57: 2387-2393, 1979.

6445 - THILL, D.C., SCHIRMAN, R.D., APPLEBY, A.P.: Influence of soil moisture, temperature, and compaction on the germination and emergence of downy brome (*Bromus tectorum*). - Weed Sci. 27: 625-630, 1979.

6446 - THILL, D.C., SCHIRMAN, R.D., APPLEBY, A.P.: Osmotic stability of mannitol and polyethylene glycol 20,000 solutions used as seed germination media. - Agron. J. 71: 105-108, 1979.

6447 - THIMANN, K.V.: Food plants and plant hormones in our future. - In: SCOTT, T.K. (ed.): Plant Regulation and World Agriculture. Pp. 1-10. Plenum Press, New York - London 1979.

6448 - THIMANN, K.V., MALIK, N., SATLER, S.: Stomatal aperture and the senescence of leaves. - In: SCOTT, T.K. (ed.): Plant Regulation and World Agriculture. Pp. 319-326. Plenum Press, New York - London 1979.

6449 - THIMANN, K.V., SATLER, S.O.: Relation between leaf senescence and stomatal closure: Senescence in light. - Proc. nat. Acad. Sci. USA 76: 2295-2298, 1979.

6450 - THIMANN, K.V., SATLER, S.: Relation between senescence and stomatal opening: Senescence in darkness. - Proc. nat. Acad. Sci. USA 76: 2770-2773, 1979.

6451 - THORNE, D.W.: Climate and crop production systems. - In: THORNE, D.W., THORNE, M.D. (ed.): Soil, Water and Crop Production. Pp. 17-29. AVI Publishing Company, Inc., Westport 1979.

6452 - THORNE, D.W.: Irrigation and crop production. - In: THORNE, D.W., THORNE, M.D. (ed.): Soil, Water and Crop Production. Pp. 96-116. AVI Publishing Company, Inc., Westport 1979.

6453 - THORNE, D.W.: Planning farms for improved production. - In: THORNE, D.W., THORNE, M.D. (ed.): Soil, Water and Crop Production. Pp. 136-150. AVI Publishing Company, Inc., Westport 1979.

6454 - THORNE, D.W.: Irrigated farming in arid and semi-arid temperate zones. - In: THORNE, D.W., THORNE, M.D. (ed.): Soil, Water and Crop Production. Pp. 229-242. AVI Publishing Company, Inc., Westport 1979.

6455 - THORNE, D.W., THORNE, M.D. (ed.): Soil, Water and Crop Production. - AVI Publishing Company, Inc., Westport 1979.

6456 - THORNE, M.D.: Soil-water-plant relations. - In: THORNE, D.W., THORNE, M.D. (ed.): Soil, Water and Crop Production. Pp. 41-53. AVI Publishing Company, Inc., Wesport 1979.

6457 - THORNE, M.D.: Physical properties of soils. - In: THORNE, D.W., THORNE, M.D. (ed.): Soil, Water and Crop Production. Pp. 70-81. AVI Publishing Company, Inc., Westport 1979.

6458 - THORNE, M.D.: Crop production systems in humid cool temperate zones. - In: THORNE, D.W., THORNE, M.D. (ed.): Soil, Water and Crop Production. Pp. 151-163. AVI Publishing Company, Inc., Westport 1979.

6459 - THORNE, M.D.: Crop production in humid warm temperate zones. - In: THORNE, D. W., THORNE, M.D. (ed.): Soil, Water and Crop Production. Pp. 183-192. AVI Publishing Company, Inc., Westport 1979.

6460 - THORPE, N., WILLMER, C.M., MILTHORPE, F.L.: Stomatal metabolism: carbon dioxide fixation and labelling patterns during stomatal movement in *Commelina cyanea*. - Aust. J. Plant Physiol. 6: 409-416, 1979.

6461 - TIBBITTS, T.W.: Humidity and plants. - BioScience 29: 358-363, 1979.

6462 - TIEN, H.T.: Photoeffects in pigmented bilayer lipid membranes. - In: BARBER, J. (ed.): Photosynthesis in Relation to Model Systems. Pp. 115-173. Elsevier, Amsterdam - New York - Oxford 1979.

*6463 - TIESZEN, L.L. (ed.): Vegetation and Production Ecology of an Alaskan Arctic Tundra. Ecophysiological Studies 29. - Springer-Verlag, New York - Heidelberg - Berlin 1978.

*6464 - TIESZEN, L.L.: Photosynthesis in the principal Barrow, , Alaska, species: A summary of field and laboratory responses. - In: TIESZEN, L.L. (ed.): Vegetation and Production Ecology of an Alaskan Arctic Tundra. Pp. 241-268. Springer-Verlag, New York - Heidelberg - Berlin 1978.

*6465 - TIESZEN, L.L., IMBAMBA, S.K.: Gas exchange of finger millet inflorescences. - Crop Sci. 18: 495-498, 1978.

6466 - TIMMER, L.W., GARNSEY, S.M., GRIMM, G.R., EL-GHOLL, N.E., SCHOULTIES, C.L.: Wilt and dieback of Mexican lime caused by *Fusarium oxysporum*. - Phytopathology 69: 730-734, 1979.

6467 - TITTEL, C.: Die Trocknung von Leguminosensamen für die Langzeitlagerung. - Kulturpflanze 27: 175-180, 1979.

6468 - TKACHUK, E.S.: Ob izmeneniyakh vodoobmena ozimoĭ pshenitsy pri razlichnykh usloviyakh vyrashchivaniya. [On changes in winter wheat water exchange under different growth conditions.] - Fiziol. Biokhim. kul't. Rast. 11: 3-10, 1979. [In R, ab: E.]

6469 - TKACHUK, E.S., SILAEVA, A.M., SHMAT'KO, I.G., BONDAR', S.F.: Ul'trastrukturnye izmeneniya khloroplastov list'ev ozimoĭ pshenitsy pri deĭstvii i posledeĭstvii pochvennoĭ zasukhi. [Ultrastructural changes in chloroplasts of winter wheat leaves during and after soil drought.] - In: Elektronnaya Mikroskopiya v Botanicheskikh Issledovaniyakh. Pp. 263-265. Zinatne, Riga 1978. [In R.]

6470 - TKACHUK, R., KUZINA, F.D.: Wheat: Relations between some physical and chemical properties. - Can. J. Plant Sci. 59: 15-20, 1979.

6471 - TOBIESSEN, P.L., SLACK, N.G., MOTT, K.A.: Carbon balance in relation to drying in four epiphytic mosses growing in different vertical ranges. - Can. J. Bot. 57: 1994-1998, 1979.

6472 - TOIVONEN, P.M.A., HOFSTRA, G.: The interaction of copper and sulphur dioxide in plant injury. - Can. J. Plant Sci. 59: 475-479, 1979.

6473 - TOMATI, U., GALLI, E.: Water stress and - SH-dependent physiological activities in young maize plants. - J. exp. Bot. 30: 557-563, 1979.

*6474 - TOMICKI, B.: A simple thermodynamic description of osmosis and imbibition. - Studia biophys. 72: 151-159, 1978.

*6475 - TOMICKI, B.: Application of a simple thermodynamic description of osmosis for calculating the coefficients σ and ω. - Studia biophys. 72: 161-167, 1978.

6476 - TOY, T.J.: Potential evapotranspiration and surface-mine rehabilitation in the Powder River Basin, Wyoming and Montana. - J. Range Manage. 32: 312-317, 1979.

6477 - TRANQUILLINI, W.: Physiological Ecology of the Alpine Timberline. Tree Existence at High Altitudes with Special Reference to the European Alps. (Ecological Studies. Vol. 31.) - Springer-Verlag, Berlin - Heidelberg - New York 1979.

6478 - TRAVIS, A.J., MANSFIELD, T.A.: Stomatal responses to light and CO_2 are dependent on KCl concentration. - Plant Cell Environ. 2: 319-323, 1979.

6479 - TRAVIS, A.J., MANSFIELD, T.A.: Reversal of the CO_2-responses of stomata by fusicoccin. - New Phytol. 83: 607-614, 1979.

*6480 - TREGUBENKO, M.Ya., FILIPPOV, G.L., VISHNEVSKIĬ, N.V.: Osobennosti vodnogo rezhima i dykhaniya razlichnykh po zasukhoustoĭchivosti gibridov kukuruzy v usloviyakh nedostatochnogo vodoobespecheniya. [Peculiarities of water regime and respiration of maize hybrids different in drought-resistance under conditions of unsufficient water supply.] - Fiziol. Biokhim. kul't. Rast. 10: 257-263, 1978. [In R, ab: E.]

6481 - TRIBOI-BLONDEL, A.-M.: Dynamique comparée de l'absorption des nitrates et de l'eau par des plantules de Blé. - C.R. Acad. Sci. Paris, Sér. D 288: 1545 - 1548, 1979.

6482 - TROMP, J.: Seasonal variations in the composition of xylem sap of apple with respect to K, Ca, Mg, and N. - Z. Pflanzenphysiol. 94: 189-194, 1979.

6483 - TROUGHTON, A.: Effect of dry soil conditions around the base of Lolium perenne plants. - In: HARLEY, J.L., SCOTT RUSSELL, R. (ed.): The Soil-Root Interface. Pp. 435. Academic Press, London - New York - San Francisco, 1979.

6484 - TROUGHTON, J.H.: $\delta^{13}C$ as an indicator of carboxylation reactions. - In: GIBBS, M., LATZKO, E. (ed.): Photosynthesis II. Pp. 140-149. Springer-Verlag, Berlin - Heidelberg - New York 1979.

*6485 - TRUKSA, J., ZÁBORSKÝ, J.: Vplyv organizácie porastu na úrodu hybridov kukurice pri závlahe. [Influence of stand organization on yield of irrigated corn hybrids.] - Rost. Výroba (Praha) 24: 1173-1182, 1978. [In Slov, ab: R,E,G.]

6486 - TSEL'NIKER, Yu.L.: Resistances to CO_2 uptake at light saturation in forest tree seedlings of different adaptation to shade. - Photosynthetica 13: 124-129, 1979.

6487 - TUCKER, C.J.: Red and photographic infrared linear combinations for monitoring vegetation. - Remote Sensing Environ. 8: 127-150, 1979.

6488 - TUCKER, C.J.: Remote Sensing of Leaf Water Content in the Near Infrared. - In: NASA technical Memorandum 80291. Pp. 1-17. Goddard Space Flight Center, Greenbelt 1979.

6489 - TUCKER, C.J., HOLBEN, B.N., ELGIN, J.H., Jr., McMURTEY, J.E. III.: The Relationship of Red and Photographic Infrared Spectral Data to Grain Yield Variation Within a Winter Wheat Field. - In: NASA technical Memorandum 80318. Pp. 1-21. Goddard Space Flight Center, Greenbelt 1979.

*6490 - TUKEY, H.B., Jr.: The effect of intermittent mist on cuttings. - Acta Hort. 79: 49-56, 1978.

6491 - TULLY, R.E., HANSON, A.D.: Amino acids translocated from turgid and water--stressed barley leaves I. Phloem exudation studies. - Plant Physiol. 64: 460-466, 1979.

6492 - TULLY, R.E., HANSON, A.D., NELSEN, C.E.: Proline accumulation in water-stressed barley leaves in relation to translocation and the nitrogen budget. - Plant Physiol. 63: 518-523, 1979.

6493 - TUR, N.S., VOROB'EV, N.V., ZHURBA, T.P.: Vliyanie izoosmoticheskikh rastvorov dekstrana, khloristogo i sernokislogo natriya na prorastanie semyan i dykhanie zarodyshcheĭ risa. [Effect of isoosmotic solutions of dextran, sodium chloride and sodium sulphate on seed germination and embryo respiration in rice.] - Fiziol. Rast. 26: 451-454, 1979. [In R.]

6494 - TURNER, N.C.: Differences in response of adaxial and abaxial stomata to environmental variables. - In: SEN, D.N., CHAWAN, D.D., BANSAL, R.P. (ed.): Structure, Function and Ecology of Stomata. Pp. 229-250. Bishen Singh Mahendra Pal Singh, Dehra Dun. 1979.

6495 - TURNER, N.C.: Drought resistance and adaptation to water deficits in crop plants. - In: MUSSEL, H., STAPLES, R. (ed.): Stress Physiology in Crop Plants. Pp. 343-372. John Wiley & Sons, Inc., New York 1979.

6496 - TYAGI, N.K., SINGH, O.P., DHRUVA NARAYANA, V.V.: Evaluation of water management systems in a tubewell irrigated farm. - Agr. Water Manage. 2: 67-78, 1979.

6497 - TYMMS, M.J., GAFF, D.F.: Proline accumulation during water stress in resurrection plants. - J. exp. Bot. 30: 165-168, 1979.

6498 - UHRSTRÖM. I., SVENSSON, S.-B.: The effect of coumarin on Young's modulus in sunflower stems and maize roots and on water permeability in potato parenchyma. - Physiol. Plant. 45: 41-44, 1979.

6499 - UMRATH, K., THALER, I., STEINER, G.: Die Wirkung von 3-Indolylessigsäure auf die Spaltöffnungsweite. - Phyton 19: 253-258, 1979.

*6500 - UNGAR, I.A.: The effects of salinity and hormonal treatments on growth and ion uptake of *Salicornia europaea*. - Bull. Soc. bot. France 125 (Actualités bot. 3/4): 95-104, 1978.

6501 - UNSWORTH, M.H.: Gaseous air pollution and productivity. - In: JOHNSON, C.B. (ed.): Physiological Processes Limiting Plant Productivity. University of Nottingham, Sutton Bonington 1979.

*6502 - UPADHYAYA, M.K., NOODÉN, L.D.: Relationship between the induction of swelling and the inhibition of elongation caused by oryzalin and colchicine in corn roots. - Plant Cell Physiol. 19: 133-138, 1978.

*6503 - URBANOVICH, T.A., MIKUL'SKAYA, S.A.: Aktyŭnasts' rĕaktsyi fotafasfarylyavannya pry roznykh rĕzhymakh vodazabespyachĕnnya raslin. [Activity of the phosphorylation reaction under different conditions of water supply to plants.] - Vestsi Akad. Navuk Belarus. SSR, Ser biyal. Navuk 1978 (2): 48-51, 1978. [In Beloruss, ab: E.]

6504 - URMANTSEV, Yu.A.: Sistemnyĭ podkhod k probleme ustoĭchivosti rasteniĭ (na primere issledovaniya zavisimosti soderzhaniya pigmentov v list'yakh fasoli ot odnovremennogo deĭstviya na nee zasukhi i zasoleniya. [Systematic approach to the problem of plant resistance (a study of pigment content in bean leaves as affected by concurently acting external drought and salinity.] - Fiziol. Rast. 26: 762-778, 1979. [In R, ab: E.]

6505 - URMANTSEV, Yu.A.: Sistemnyĭ podkhod k probleme ustoĭchivosti rasteniĭ (adekvatnost' i interpretatsiya regressionnykh uravneniĭ zavisimosti soderzhaniya pigmentov v list'yakh fasoli ot odnovremennogo deĭstviya na nee zasukhi i NaCl. [Systematic approach to the problem of plant resistance (adequacy and interpretation of the regression equations of dependence of bean leaf pigment content on external drought and NaCl acted simultaneously).] - Fiziol. Rast. 26: 1233-1244, 1979. [In R, ab: E.]

*6506 - USHIJIMA, T., TAZAKI, T.: Photosynthetic activity and water metabolism in some higher plants. - In: MONSI, M., SAEKI, T. (ed.): Ecophysiology of Photosynthetic Productivity. JIBP Synthesis. Vol. 19. Pp. 37-46. University of Tokyo Press, Tokyo 1978.

*6507 - VAN ANDEL, J., BOS, W., ERNST, W.: An experimental study on two populations of *Chamaenerion angustifolium* (L.) Scop. (= *Epilobium angustifolium* L.) occurring on contrasting soils, with particular reference to the response to bicarbonate. - New Phytol. 81: 763-772, 1978.

6508 - Van ASSCHE, F., CLIJSTERS, H., MARCELLE, R.: Photosynthesis in *Phaseolus vulgaris* L., as influenced by supra-optimal zinc nutrition. - In: MARCELLE, R., CLIJSTERS, H., Van POUCKE, M. (ed.): Photosynthesis and Plant Development. Pp. 175-184. Dr. W. Junk bv Publishers, The Hague - Boston - London 1979.

6509 - VAN den DRIESSCHE, R., CHEUNG, K.-W.: Relationship of stem electrical impedance and water potential of Douglas-fir seedlings to survival after cold storage. - Forest. Sci. 25: 507-517, 1979.

6510 - VAN DORP, F., SINNAEVE, J.: Simulation of changes in nutrient concentrations in the root environment. II. Models of short term changes around the roots. - In: HARLEY, J.L., SCOTT RUSSELL, R. (ed.): The Soil-Root Interface. Pp. 437. Academic Press, London - New York - San Francisco 1979.

6511 - VAN DORP, F., SINNAEVE, J., FRISSEL, M.J.: Simulation of changes in nutrient concentrations in the root environment. I. Long term models based on experiments with gravel culture. - In: HARLEY, J.L., SCOTT RUSSELL, R. (ed.): The Soil-Root Interface. Pp. 436-437. Academic Press, London - New York - San Francisco 1979.

*6512 - VAN NOORDWIJK, M.: Zoutophoping en beworteling bij de teelt van tomaten op steenwol. [Distribution of salts and root development in the culture of tomatoes on rock wool.] - Inst. Bodemvruchtbaarheid Haren-Gr. Rapp. 3-78: 1-21, 1978. [In Dutch, ab: E.]

6513 - VAN OORSCHOT, J.L.P., VAN LEEUWEN, P.H.: Recovery from inhibition of photosynthesis by metamitron in various plant species. - Weed Res. 19: 63-67, 1979.

*6514 - VARFOLOMEEV, L.A.: Vlazhnost' zavyadaniya rasteniĭ na dvukhkomponentnykh sme-
         syakh bolotno-podzolistoĭ pochvy. [Wilting moisture of two-component mixtures
         from upper horizonts of a swamp-podzolic soil.] - Pochvovedenie 1978 (10): 45-
         52, 1978. [In R, ab: E.]

6515 - VASIĆ, G.: Režim navodnjavanja kukuruza i količina vode koju potroši u eko-
        logiškim uslovima Zemunskog polja. [Maize irrigation system and amount of
        water consumed under the ecological conditions of Zemun polje.] - Arch. poljo-
        privredne Nauke 32 (118): 23-44, 1979. [In Croat, ab: E.]

*6516 - VERMA, B., RAMANATH, B., HANUMANTHAPPA, B.: Use of run-off water for supple-
         mental irrigation of rabi sorghum in deep black soils of Karnataka. - Ind. J.
         Agron. 23: 219-229, 1978.

*6517 - VESTER, G., CRAWFORD, R.M.M.: Verschiedene Provenienzen von *Pinus concorta*
         Loudon und Überflutungsstress: Klassifikation auf Grund morphologischer und
         metabolischer Kriterien. - Flora 167: 433-444, 1978.

*6518 - VIEIRA da SILVA, J.: Quelques similitudes entre les agressions provoquées par
         la salinité et celles causées par la sécheresse et par le gel. - Bull. Soc.
         bot. France 125 (Actualités bot. 3/4): 275-284, 1978.

6519 - VIGNES, D., PLANCHON, C.: Structure, eclairement et echanges gazeux d'une cul-
        ture de Soja (*Glycine max* L. Merr.) - Photosynthetica 13: 136-145, 1979.

*6520 - VITKOV, M., DEKOV, D.: Prouchvane v"rkhu toreneto i napoyavaneto na polskiya
         grakh na opodzolen chernozem. [Study on the mineral fertilization and irriga-
         tion of field pea on podzolized chernozem soil.] - Rasteniev"dni Nauki 15(5):
         71-76, 1978. [In Bulg., ab: R,E.]

*6521 - VITKOV, M., PAVLOVA, S.: Vliyanie na napoyavaneto i toreneto v"rkhu vodniya
         potentsial i koncentratsiyata na klet"chniya sok pri fasula. [Water potential
         and cell sap concentration in beans as affected by irrigation and fertiliz-
         ation.] - Rasteniev"dni Nauki 15 (9-10): 32-35, 1978. [In Bulg., ab: R,E.]

6522 - VOLKOV, V.Ya., VELIKANOV, G.A.: Izuchenie transporta vody v membranakh kletok
        drozhzheĭ impul'snym metodom yadernogo magnitnogo rezonansa. [NMR studies of
        water transport in yeast cell membranes.]- Biofizika 24: 77-81, 1979.[In R,ab: E.]

*6523 - WACQUANT, J.-P., PASSAMA, L.: Proportions d'ions K$^+$ dans la plante et résis-
         tance au sel de divers halophytes: relation avec les propriétés d'adsorption
         des racines. - Bull. Soc. bot. France 125 (Actualités bot. 3/4): 111-121, 1978.

*6524 - WAHUA, T.A.T., MILLER, D.A.: Leaf water potentials and light transmission of
         intercropped sorghum and soybeans. - Exp. Agr. 14: 373-380, 1978.

6525 - WAIN. R.L.: Potential for regulation of plant growth and development. - In:
        SCOTT, T.K. (ed.): Plant Regulation and World Agriculture. Pp. 155-164. Plenum
        Press, New York - London 1979.

6526 - WALKER, G.K., HATFIELD, J.L.: Test of the stress-degree-day concept using mul-
        tiple planting dates of red kidney beans. - Agron. J. 71: 967-971, 1979.

6527 - WALKER, R.R., KRIEDEMANN, P.E., MAGGS, D.H.: Growth, leaf physiology and fruit
        development in salt-stressed guavas. - Aust. J. agr. Res. 30: 477-488, 1979.

6528 - WALLENDER, W.W., GRIMES, D.W., HENDERSON, D.W., STROMBERG, L.K.: Estimating
        the contribution of a perched water table to the seasonal evapotranspiration
        of cotton. - Agron. J. 71: 1056-1060, 1979.

6529 - WARDLE, K., QUINLAN, A., SIMPKINS, I.: Abscisic acid and the regulation of
        water loss in plantlets of *Brassica oleracea* L. var. *botrytis* regenerated
        through apical meristem culture. - Ann. Bot. 43: 745-752, 1979.

6530 - WARDLE, K., SIMPKINS, I.: Stomatal responses of *Phaseolus vulgaris* L. seedlings to potassium chloride in the nutrient solution - J. exp. Bot. 30: 1195-1200, 1979.

6531 - WAREING, P.F.: Temperature responses and yield in temperate crops. - In: SCOTT, T.K. (ed.): Plant Regulation and World Agriculture. Pp. 129-139. Plenum Press, New York - London 1979.

6532 - WARING, R.H., ROBERTS, J.M.: Estimating water flux through stems of Scots pine with tritiated water and phosphorus-32. - J. exp. Bot. 30: 459-471, 1979.

*6533 - WARING, R.H., RUNNING, S.W.: Sapwood water storage: its contribution to transpiration and effect upon water conductance through the stems of old-growth Douglas-fir. - Plant Cell Environ 1: 131-140, 1978.

6534 - WARRICK, A.W., AMOOZEGAR-FARD, A., LOMEN, D.O.: Linearized moisture flow from line sources with water extraction. - Trans. ASAE 22: 549-553, 559, 1979.

*6535 - WATT, L.A.: Some characteristics of the germination of Queensland blue grass on cracking black earths. - Aust. J. agr. Res. 29: 1147-1155, 1978.

6536 - WEARING, A.H., BURGESS, L.W.: Water potential and the saprophytic growth of *Fusarium roseum* "Graminearum". - Soil Biol. Biochem 11: 661-667, 1979.

6537 - WEATHERLEY, P.E.: The hydraulic resistance of the soil-root interface - A cause of water stress in plants. - In: HARLEY, J.L., SCOTT RUSSELL, R. (ed.): The Soil-Root Interface. Pp. 275-286. Academic Press, London - New York - San Francisco 1979.

6538 - WEATHERLY, A.B., DANE, J.H.: Effect of tillage on soil-water movement during corn growth. - Soil Sci. Soc. Amer. J. 43: 1222-1225, 1979.

*6539 - WEIHE, K. von: Untersuchungen zur Ökologie von *Festuca rubra* L. ssp. *litoralis* (G.F.W. Meyer) Auquier (Temperatur und Meersalzwirkung). - Beitr. Biol. Pflanz. 54: 125-143, 1978.

*6540 - WEIHE, K. von: Untersuchungen zur Ökologie von *Puccinellia maritima* (Huds.) Parl. (Temperatur und Meersalzwirkung). - Beitr. Biol. Pflanz. 54: 145-163, 1978.

6541 - WEILAND, R.T., STUTTE, C.A.: Pyro-chemiluminiscent differentiation of oxidized and reduced N forms evolved from plant foliage. - Crop Sci. 19: 545-547, 1979.

*6542 - WENT, F.W., BABU, V.R.: The effect of dew on plant water balance in *Citrullus vulgaris* and *Cucumis melo*. - Physiol. Plant. 44: 307-311, 1978.

6543 - WERKER, E., MARBACH, I., MAYER, A.M.: Relation between the anatomy of the testa, water permeability and the presence of phenolics in the genus *Pisum*. - Ann. Bot. 43: 765-771, 1979.

6544 - WEST, D.W., MERRIGAN, I.F., TAYLOR, J.A., COLLINS, G.M.: Soil salinity gradients and growth of tomato plants under drip irrigation. - Soil Sci. 127: 281-291, 1979.

6545 - WEST, G.C., SALO, A.C.: Seasonal changes in water and fat content and fatty acid composition of the catkin buds of Alaska willow (*Salix alaxensis*). - Oecologia 41: 207-218, 1979.

6546 - WESTE, G., VITHANAGE, K.: Survival of chlamydospores of *Phytophthora cinnamomi* in several non-sterile, host-free forest soils and gravels at different soil water potentials. - Aust. J. Bot. 27: 1-9, 1979.

6547 - WEYERS, J.D.B., HILLMAN, J.R.: Uptake and distribution of abscisic acid in *Commelina* leaf epidermis. - Planta 144: 167-172, 1979.

6548 - WEYERS, J.D.B., HILLMAN, J.R.: Sensitivity of *Commelina* stomata to abscisic acid. - Planta 146: 623-628, 1979.

6549 - WHATLEY, J.M.: Plastid development in the primary leaf of *Phaseolus vulgaris*: variations between different types of cell. - New Phytol. 82: 1-10, 1979.

6550 - WHIGHAM, D.K., STOLLER, E.W.: Soybean desiccation by paraquat, glyphosate, and ametryn to accelerate harvest. - Agron J. 71: 630-633, 1979.

6551 - WHITCOMB, C.E.: Effects of black plastic and mulches on growth and survival of landscape plants. - Oklahoma agr. Exp. Sta. Res. Rep. P-791: 8-11, 1979.

6552 - WIEBE, H.H., BROWN, R.W.: Temperature gradient effects on in situ hygrometer measurements of soil water potential. II. Water movement. - Agron. J. 71: 397 -401, 1979.

6553 - WIEGAND, C.L., RICHARDSON, A.J., KANEMASU, E.T.: Leaf area index estimates for wheat from LANDSAT and their implications for evapotranspiration and crop modeling. - Agron. J. 71: 336-342, 1979.

6554 - WIGHTMAN, F.: Modern chromatographic methods for the identification and quantification of plant growth regulators and their application to studies of the changes in hormonal substances in winter wheat during acclimation to cold stress conditions. - In: SCOTT, T.K. (ed.): Plant Regulation and World Agriculture. Pp. 327-377. Plenum Press, New York - London 1979.

6555 - WILCOX, J.C.: Some effects of use of transparent shields over small evaporimeters. - Can. J. Plant Sci. 59: 445-451, 1979.

6556 - WILD, A.: Physiologie der Photosynthese Höherer Pflanzen. Die Anpassung an die Lichtbedingungen. - Ber. Deut. bot. Ges. 92: 341-364, 1979.

6557 - WILLEMOT, C., PELLETIER, L.: Effect of drought on frost resistance and fatty acid content of young winter wheat plants. - Can. J. Plant Sci. 59: 639-643, 1979.

6558 - WILLERT, D.J., von: Vorkommen und Regulation des CAM bei Mittagsblumengewächsen (*Mesembryanthemaceae*). - Ber. Deut. bot. Ges. 92: 133-144, 1979.

6559 - WILLERT, D.J., von, BRINCKMANN, E., SCHEITLER, B., THOMAS, D.A., TREICHEL, S.: The activity and malate inhibition/stimulation of phosphoenolpyruvate-carboxylase in crassulacean-acid-metabolism plants in their natural environment. - Planta 147: 31-36, 1979.

6560 - WILLIAMS, R.J., BROERSMA, K., VAN RYSWYK, A.L.: The effects of nitrogen fertilization on water use by crested wheatgrass. - J. Range Manage. 32: 98-100, 1979.

*6561 - WILLIAMS, T.H.L.: An automatic scanning and recording tensiometer system - J. Hydrol. 39: 175-183, 1978.

6562 - WILLMER, C.M., SEXTON, R.: Stomata and plasmodesmata. - Protoplasma 100: 113-124, 1979.

6563 - WILSON, J.R., FISHER, M.J., SCHULZE, E.-D., DOLBY, G.R., LUDLOW, M.M.: Comparison between pressure-volume and dewpoint-hygrometry techniques for determining the water relations characteristics of grass and legume leaves. - Oecologia 41: 77-88, 1979.

*6564 - WILSON, R.A.: Root crops. - In: ALVIM, P. de T., KOZLOWSKI, T.T. (ed.): Ecophysiology of Tropical Crops. Pp. 187-236. Academic Press, New York - San Francisco - London 1977.

6565 - WILSON, R.G., Jr.: Germination and seedling development of Canada thistle (*Cirsium arvense*). - Weed Sci. 27: 146-151, 1979.

6566 - WINTER, K.: Effect of different CO_2 regimes on the induction of crassulacean acid metabolism in *Mesembryanthemum crysallinum* L. - Aust. J. Plant Physiol. 6: 589-594, 1979.

*6567 - WIT, C.T. de, et al.: Simulation of Assimilation, Respiration and Transpiration of Crops. - Pudoc, Wageningen 1978.

6568 - WITHERS, J.R.: Studies on the status of unburnt *Eucalyptus* woodland at Ocean Grove, Victoria. V. The interactive effects of droughting and shading on seedlings under competition. - Aust. J. Bot. 27: 285-300, 1979.

6569 - WITHERS, N.J.: Effects of water stress on *Lupinus albus* I. Response of vegetative growth to water stress during a single growth stage at two humidity levels. - N. Zeal. J. agr. Res. 22: 445-454, 1979.

6570 - WITHERS, N.J.: Effects of water stress on *Lupinus albus* II. Response of seed yield to water stress during a single growth stage at two humidity levels. - N. Zeal. J. agr. Res. 22: 455-461, 1979.

6571 - WITHERS, N.J., EDGE, E.A.: Effects of water stress on *Lupinus albus* IV. Response to high temperature and adequate and restricted water. - N. Zeal. J. agr. Res. 22: 571-575, 1979.

6572 - WITHERS, N.J., FORDE, B.J.: Effects of water stress on *Lupinus albus* III. Response of seed yield and vegetative growth to water stress imposed during two or three growth stages. - N. Zeal. J. agr. Res. 22: 463-474, 1979.

6573 - WITHERS, N.J., FORDE, B.J.: Translocation of ^{14}C in *Lupinus albus*. - N. Zeal. J. agr. Res. 22: 561-569, 1979.

6574 - WITTWER, S.H.: Agricultural production - research imperatives for the future. In: SCOTT, T.K. (ed.): Plant Regulation and World Agriculture. Pp. 11-33. Plenum Press, New York - London 1979.

6575 - WOLEDGE, J.: Effect of flowering on the photosynthetic capacity of ryegrass leaves grown with and without natural shading. - Ann. Bot. 44: 197-207, 1979.

6576 - WONG, S.C.: Elevated atmospheric partial pressure of CO_2 and plant growth I. Interactions of nitrogen nutrition and photosynthetic capacity in C_3 and C_4 plants. - Oecologia 44: 68-74, 1979.

6577 - WONG, S.C., COWAN, I.R., FARQUHAR, G.D.: Stomatal conductance correlates with photosynthetic capacity. - Nature 282: 424-426, 1979.

6578 - WOODWARD, F.I., YAQUB, M.: Integrator and sensors for measuring photosynthetically active radiation and temperature in the field. - J. appl. Ecol. 16: 545-552, 1979.

6579 - WRIGHT, D., HEBBLETHWAITE, P.D.: Lodging studies in *Lolium perenne* grown for seed. 3. Chemical control of lodging. - J. agr. Sci. 93: 669-679, 1979.

6580 - WRIGHT, S.T.C.: The effect of 6-benzyladenine and leaf ageing treatment on the levels of stress-induced ethylene emanating from wilted wheat leaves. - Planta 144: 179-188, 1979.

*6581 - WRISCHER, M.: Ultrastructural localization of diaminobenzidine photooxidation in etiochloroplasts. - Protoplasma 97: 85-92, 1978.

6582 - WYN JONES, R.G., BRADY, C.J., SPEIRS, J.: Ionic and osmotic relations in plant cells. - In: LAIDMAN, D.L., WYN JONES, R.G. (ed.): Recent Advances in the Biochemistry of Cereals. Pp. 63-103. Academic Press, London - New York - San Francisco 1979.

6583 - YAMAMOTO, T., WATANABE, S., HARADA, H.: Studies on leaf burn of pear trees. IX. Analysis of promoting factors in development of leaf burn: Relationships between leaf burn, daily courses of meteorological variables, and leaf water deficits. - Bull. Yamagata Univ., Agr. Sci. 8(2): 31-46, 1979.

6584 - YI, C., TODD, G.W.: Changes in ribonuclease activity of wheat plants during water stress. - Physiol. Plant. 46: 13-18, 1979.

*6585 - YOPP, J.H., MILLER, D.M., TINDALL, D.R., BREGGER, T.: Environmental, metabolic and developmental aspects of osmoregulation in the halophilic blue-green alga, *Aphanothece halophytica*. - Plant Physiol. 61 (Suppl.): 95, 1978.

6586 - YOSHIDA, T.: Relationship between stomatal frequency and photosynthesis in barley. - JARQ 13: 101-105, 1979.

6587 - YOUNG, D.R., SMITH, W.K.: Influence of sunflecks on the temperature and water relations of two subalpine understory congeners. - Oecologia 43: 195-205, 1979.

6588 - YOUNG, E., OLCOTT-REID, B.: Siberian C rootstock delays bloom of peach. - J. Amer. Soc. hort. Sci. 104: 178-181, 1979.

*6589 - YOUNIS, H.M., BOYER, J.S., GOVINDJEE: Conformation and activity of chloroplast coupling factor exposed to low leaf water potentials. - Plant Physiol. 61 (Suppl.): 77, 1978.

6590 - YOUNIS, H.M., BOYER, J.S., GOVINDJEE: Conformation and activity of chloroplast coupling factor exposed to low chemical potential of water in cells. - Biochim. biophys. Acta 548: 328-340, 1979.

6591 - ZACCAI, G., GILMORE, D.J.: Areas of hydration in the purple membrane of *Halobacterium halobium:* A neutron diffraction study. - J. mol. Biol. 132: 181-191, 1979.

6592 - ZAGDAŃSKA, B., PACANOWSKA, A.: Dehydration tolerance of spring wheat and its relation to plant growth and productivity under soil drought conditions. - Biol. Plant 21: 452-461, 1979.

6593 - ZAGDAŃSKA, B., PACANOWSKA, A.: Dehydration tolerance in spring wheat seeds. - Biol. Plant. 21: 462-467, 1979.

*6594 - ZANGIACOMI, L., BONISCHOT, R., GUCKERT, A.: Influence de la variation du régime hydrique sur la flore et la productivité des prairies permanentes de la vallée de l'Yron (Meuse). - Bull. E.N.S.A.I.A. (Nancy) 20: 47-54, 1978.

6595 - ZAROGIANNIS, V.: Beregnung und Standraum bei Mais (*Zea mays* L.). - Bodenkultur 30: 281-303, 1979.

6596 - ZARRA, I., MASUDA, Y.: Growth and cell wall changes in rice coleoptiles growing under different conditions I. Changes in turgor pressure and cell wall polysaccharides during intact growth. - Plant Cell Physiol. 20: 1117-1124, 1979.

6597 - ZAVITKOVSKI, J.: Energy production in irrigated, intensively cultured plantations of *Populus* "Tristis #1" and Jack Pine. - Forest Sci. 25: 383-392, 1979.

6598 - ZAVITKOVSKI, J., HANSEN, E.A., McNEEL, H.A.: Nitrogen-fixing species in short rotation systems for fiber and energy production. - In: GORDON, J.C., WHEELER, C.T., PERRY, D.A. (ed.): Symbiotic Nitrogen Fixation in the Management of Temperate Forests. Pp. 388-402. Oregon State University, Corvallis 1979.

6599 - ZEEVAART, J.A.D.: Chemical and biological aspects of abscisic acid. - In: MANDAVA, N.B. (ed.): Plant Growth Substances. Pp. 99-114. American Chemical Society, Washington 1979.

6600 - ZEIGER, E., HEPLER, P.K.: Blue light-induced, intrinsic vacuolar fluorescence in onion guard cells. - J. Cell Sci. 37: 1-10, 1979.

6601 - ZHELEV, R., PAVLOV, P.A.: Razrastvane na korenovata sistema na tritikale v zavisimost ot pochveneta vlaga. [Development of root system of *Triticale* in dependence on soil moisture.] - Rasteniev"dni Nauki 16(3): 8-12, 1979. [In Bulg., ab: R,E.]

6602 - ZIMMERMANN, U., HÜSKEN, D.: Theoretical and experimental exclusion of errors in the determination of the elasticity and water transport parameters of plant cells by the pressure probe technique. - Plant Physiol. 64: 18-24, 1979.

6603 - ZINSOU, C., SCHOCH, P.-G.: Mise en évidence de la participation des feuilles adultes à l'expression de l'indice stomatique de la jeune feuille en différenciation du *Vigna sinensis* L. - Physiol. vég. 17: 327-336, 1979.

*6604 - ZÓBEL, D.B., LIN, T.-P., LIU, V.T.: Stomatal distribution on leaves of three species of *Chamaecyparis*. - Taiwania 23: 1-6, 1978.

6605 - ZYALALOV, A.A.: O retsirkulyatsii kaliya v steble v svyazi s transportom vody. [Recycling of potassium in the stem in relation to water transport.] - Fiziol. Rast. 26: 579-583, 1979. [In R, ab: E.]

AUTHORS' INDEX

(Cumulative index to volumes 1 - 5)

Authors' names are presented in the form in which they appear in the respective publication. The names from papers published in Cyrillic character are transcribed as shown in Instructions for Use. Alternative spelling and form of the name of the same author are usually cross-indexed.

A

AARSTAD, J.S. 1846
AASE, J.K. 1583
ABATUROV, Yu.D. 619
ABAWI, G.S. 980, 1460, 4754
ABAZA, M. 2940
ABDEL-AL, Z.E. 981, 982
ABDELHAFEEZ, A.T. 1, see also
ABDEL HAFEEZ, A.T. 983
ABD EL RAHMAN, A.A. 984
ABDEL-WAHID, A.A. 3668
ABDULRAHMAN, F.S. 3687
ABDURAKHMANOV, A.A. 3688
ABEL, G.H. 985
ABISALOV, R.S. 1835, 4901
ABO, F. 4233, 4234
ABOU-RAYA, M.A. 2, 1348, 4066
ABRAHAM, J. 2161
ABRAMOVITZ, R. 6436
ABROL, I.P. 2813
ABU-SHAKRA, S. 3736
ACEVEDO, E. 1367, 1562, 1563, 4089
ACEVES-N., E. 986, 987, 2480, see also
ACEVES-NAVARRO, E. 614
ACHARYYA, N. 3586
ACKERSON, R.C. 3, 2481, 2482, 2483, 2484
ACOCK, B. 4, 988, 5249
ADAMS, A.F.R. 1509
ADAMS, J.A. 2485, 3689
ADAMS, J.E. 989, 3690
ADAMS, M. 201
ADAMS, M.S. 587, 5545, 5926
ADAMS, P.B. 990
ADAMS, R.E. 5
ADAMSKI, T. 1051
ADAMSON, R.M. 2517
ADDINK, J.W. 6
ADEDIPE, N.O. 991, 992, 5062, 5250
ADJAHOSSOU, F. 5251
ADJEI-TWUM, D.C. 993
ADRIANO, D.C. 3295
AEROV, I.L. 3691

AFSCHAR, I. 2818
AFSHAR, A. 3692
AGABBIO, M. 3693
AGARWAL, M.C. 3609
AGARWAL, S. 1476
AGARWAL, S.K. 2486, 5252
AGATA, W. 3552, 5056
AGGARWAL, G.C. 994, 3694
AGNIHOTRI, R.C. 6220
AGRAWAL, P.K. 5253
AHARONI, N. 7, 995, 2487, 3695, 3696, 3697
AHLGRIMM, H.-J. 2488, 5254
AHMAD, I. 996, 2489, 5255
AHMAD, K.J. 8, 5256, 5257, 5258
AHMAD, N. 3664
AHMED, A.M. 4069, 5259, 5260
AHMED, J. 1070
AHMED, M.el H. 4240
AHO, N. 2490, 5261
AHRING, R.M. 4409
AHUJA, L.R. 9
AIKIN, W.J. 10
AIMI, R. 11
AJTAY, G.L. 5659
AKALEHIYWOT, T. 3698, 3699
AKHMEDOV, G.S. 997
AKHMEDOV, I.S. 5970
AKHMEDOV, N. 706
AKHRAMENKO, L.P. 1471
AKHTAR, P. 3700
AKINLEYE, D. 4834
AKINSOROTAN, A.O. 1183
AKITA, S. 3550, 5262, 5795
AKOPYAN, G.A. 998
AKSENOV, S.I. 12, 999, 2491, 2492, 3201, 3701, 3702, 3727
AKSENOVICH, A.V. 1000
AKSYONOV, S.I. 2493
ALBERDA, T. 415, 3703
ALBERGONI, F.G. 5263
ALBERT, G. 736
ALBERT, R. 3704

BRETTHAUER, E.W.   565
BREWER, W.L.J.   2725
BREY, W.S.   2120
BREZEANU, A.   1146
BRIDGEN, M.P.   1147
BRIDGWATER, F.E.   5401
BRIEN, C.J.   503
BRIENS, M.   5369
BRIGGS, K.G.   6397
BRIGGS, R.E.   3251
BRINCKMANN, E.   5370, 6559
BRINKHUIS, B.H.   1148, 5371
BRINKMAN, M.A.   5372
BRINKMANN, K.   1149
BRISKE, D.D.   2445
BRISSON, J.D.   732, 3352, 6251, 6252
BRIX, H.   5373
BROADBENT, F.E.   744
BROCK, T.D.   95, 96, 1150, 3173, 5912
BROCKLEHURST, P.A.   3844
BROCKWAY, C.E.   6253
BROCKWELL, J.   2590
BRODY, S.   5567
BROERSMA, K.   6560
BROGÅRDH, T.   97, 98, 387, 1151, 1528, 1625, 4416, 5781
BRONCHART, R.   1259
BROOK, P.J.   3845
BROOKES, P.C.   5374, 5375
BROSZ, D.D.   2702
BROUÉ, P.   2591
BROUWER, R.   99
BROWER, D.L.   1152
BROWN, A.D.   1153, 3830, 3846, 3847, 3848, 4055, 5376, 5912
BROWN, D.C.W.   2726, 5377
BROWN, D.H.   52, 5378, 5384
BROWN, D.M.   6320
BROWN, F.A., Jr.   1154, 5379
BROWN, G.N.   2583, 3849, 6012
BROWN, J.F.   1693, 1694
BROWN, J.H., Jr.   100
BROWN, K.W.   391, 1155, 2328, 3850, 4944, 5380
BROWN, L.C.   4819, 5115
BROWN, L.F.   2592, 2593
BROWN, L.M.   3851, 3852, 3853, 3854
BROWN, M.J.   2594
BROWN, R.H.   1086, 1087, 1088, 2551, 3822, 3855, 3856, 5381
BROWN, R.T.   145
BROWN, R.W.   1156, 1157, 2595, 2954, 3641, 5173, 6552
BROWN, T.H.   2596
BROWNE, C.L.   2579, 3857, 3858, 3859
BROWNING, G.   101, 102, 2598, 5596
BROWNING, V.D.   3158, 3860
BRUCE, R.R.   1158, 3196, 3227

BRUCK, K.   5731
BRUCKERHOFF, D.N.   338, 340
BRÜGGEMANN, M.   3861
BRUINSMA, J.   5382
BRUN, L.J.   103
BRUN, W.A.   136, 861, 3924, 5383
BRUNKE, H.   4837
BRUNNER, U.   1159, 2599, 3862
BRUSEWITZ, G.H.   158
BRYAN, H.H.   1160
BRYAN, W.E.   244, 1391
BRYUKVIN, V.G.   2733, 3055
BRZOZOWSKA, J.   311, 2600
BUCHANAN, B.A.   2601
BUCHANAN, D.W.   1161, 6441
BUCHSBAUM, S.   2339
BUCHWALD, T.   1049
BUCK, G.W.   5378, 5384
BUCKNER, E.   2602
BUCKNER, R.C.   5385
BUCKS, D.A.   5643
BUGAKOVA, A.N.   104
BUGGELN, R.G.   3863
BUGLOS, J.   4545
BUICULESCU, I.   5386
BUI HUY THIEN   4175
BUIJTENEN, J.P., van   see VAN BUIJTENEN, J.P.
BUJTÀS, K.   3965
BUKHAR, I.E.   5387
BUKHOV, N.G.   5808
BUKOVAC, M.J.   1163, 1378, 2850, 4914, 5293
BULA, R.J.   2924
BULARD, C.   2534
BULATOVA, T.A.   3724
BULL, T.A.   1164
BULTOT, F.   1165, 1166
BUMBIERIS, M.   1167
BUNCE, J.   2414, 3634
BUNCE, J.A.   1168, 1169, 2603, 2604, 2605, 2606, 3245, 3864, 3865, 3866, 5388
BUNEA, A.   832
BÜNEMANN, G.   302
BUNNENBERG, C.   3867
BUNTING, A.H.   1170
BUNTLEY, G.J.   244
BURBANO, J.L.   1171, 2012
BURBÖCK, W.   1172
BURCH, G.J.   3868, 3869, 5389
BURDASOV, V.M.   1173
BURDEN, R.S.   2607
BURDON, J.J.   118
BUREŠ, V.   3870
BURGDORF, O.   1174
BURGE, M.N.   2608
BURGESS, L.W.   6536
BURGESS, M.D.   105
BURGHARDT, W.   1175, 1176
BURGHARDTOVÁ, K.   1177
BURK, J.H.   3871

CHLADOVÁ, J.   3087, 4537
CHLOUPEK, O.   1221
CHOLICK, F.A.   2650
CHONAN, N.   1656, 5427
CHONG, C.   5428
CHOONG, S.A.   3916
CHOPRA, H.K.   4520
CHOUBEY, S.D.   3318, 3427
CHOUDHURI, M.A.   1100, 1101
CHOUDHURY, N.K.   1222, 1276
CHOW, C.S.   1154
CHOW, T.L.   2651
CHOWDHARY, R.K.   1960
CHRIST, R.   5102
CHRIST, R.A.   3917
CHRISTEN, A.A.   1242
CHRISTENSEN, C.M.   1223
CHRISTENSEN, O.   1224, 3918
CHRISTENSEN, P.   1225
CHRISTERSSON, L.   2652
CHRISTIANSEN, J.E.   2653
CHRISTIANSEN, M.N.   3919, 5429
CHRISTIE, A.J.R.   3038
CHRISTIE, E.K.   133, 134, 3920
CHRISTY, A.L.   231, 2654
CHU, A.C.P.   2655, 2656, 3133,
3921, 5430
CHU, C.C.   495
CHU, M.C.   1309
CHU, T.M.   1226, 1227
CHUBEY, B.B.   2717, 3922
CHUDINA, V.I.   135, 4425
CHUNG, H.H.   1228
CHURCHILL, D.M.   1395
CHUROVÁ, K.   3923
ČIAMPOROVÁ, M.   1229, 1230
CIHA, A.J.   136, 3924
CIMANOWSKI, J.   1231
CIOTEA, V.   3466
CLARK, B.J.   2657, 3925
CLARK, C.   5774
CLARK, J.A.   70, 1619
CLARK, J.B.   137
CLARK, J.F.   3926
CLARK, L.E.   3812
CLARK, R.A.   1082, 1912
CLARKE, A.R.P.   3391
CLARKE, J.M.   3927, 3928,
5431
CLARKSON, D.T.   1232, 2106,
2130, 3929, 6250
CLARKSON, N.M.   1233
CLARY, W.P.   1234
CLAUS, S.   2658
CLAY, K.   3930
CLAYTON, R.K.   3931
CLEGG, B.   5868
CLEGG, B.R.   3443
CLEGG, J.S.   5450
CLELAND, R.   5125
CLELAND, R.E.   138, 175, 176,
3932
CLEMENS, J.   2659, 3933, 3934

CLIJSTERS, H.   5117, 6508
CLINE, J.F.   2660
CLINE, R.G.   1235
CLINGENPEEL, W.J.   88
CLOUGH, B.F.   139
CLOUGH, J.M.   3935, 5432, 5433
CLUCAS, R.D.   5434
CLUTTERBUCK, B.J.   3936
COBB, A.H.   3937
COBB, F.W., Jr.   4164
COCHRANE, J.   2004, 2005
COCK, J.H.   5930
COCKBURN, W.   5435, 5441
COCKROFT, B.   726
COCOZA TALIA, M.   3938
COCUCCI, M.   1236
COCUCCI, S.   1236
CODACCIONI, M.   4455
CODRON, H.   3939
COE, M.J.   1237
COGELS, O.   1238
COHEN, D.   58, 1423
COHEN, W.S.   3940
COHEN, Y.   1096, 2661, 3059
COHN, M.A.   1239, 3941
COKE, L.B.   140, 852
COLE, C.V.   2650
COLE, N.H.A.   2662
COLE, P.J.   3942
COLEMAN, D.C.   6145
COLEMAN, R.A.   1394
COLESANTI, F.   4550
COLHOUN, J.   2413
COLLATZ, G.J.   2663, 3160, 4606
COLLATZ, J.   1240
COLLINS, D.   3943
COLLINS, G.M.   6544
COLLINS, M.T.   3121
COLLINS, N.J.   1927, 5397
COLLINS, O.D.G.   2664
COLLINS, W.J.   2506
COLLINSON, M.E.   1212
COLLIS-GEORGE, N.   141, 3944
COLMAN, B.   289, 3945, 4560,
5436
COLMAN, R.L.   142
COLWICK, R.F.   334
COMBRES, J.C.   5608
COMPTON, W.A.   3267
CONDE, L.F.   143
CONESA, A.P.   5437
CONGLY, H.   1241
CONNOR, D.J.   144, 908, 2665
CONOVER, C.A.   3946
CONSIDINE, J.A.   1502
CONSTABLE, G.A.   3947
CONSTANT, R.   1907
CONSTANTIN, R.J.   145
CONSTANTINI, A.   3948
CONTOUR-ANSEL, D.   5438
COOK, A.A.   6405
COOK, E.R.   2666
COOK, R.J.   1242

FRANGMEIER, D.D. 2792
FRANICH, R.A. 2793
FRANK, A.B. 1387, 2794, 2795,
4120
FRANKE, G. 5612
FRANQUIN, P. 2796, 5613
FRASIER, G.W. 1244
FRATICELLI, A. 5090
FREARSON, E.M. 1284
FREEBERG, L.R. 5614
FREEMAN, B. 5615
FREEMAN, B.M. 1388, 2797,
5616
FREEMAN, D.C. 1389
FREEMAN, E.A. 3772
FREEMAN, T.P. 191, 243, 1325,
1326, 4046
FREESE, F. 2048
FREIJSEN, A.H.J. 4867
FRENCH, B.K. 3985
FRENCH, D.W. 1080
FRENCH, J.C. 1390
FRENCH, N.R. 5617
FRENEY, J.R. 1298
FRENKEL, H. 4364
FRENYÓ, V. 5618
FRETZ, T.A. 4554
FREY, K.J. 5372
FRIBOURG, H.A. 244, 1391
FRIEDMAN, I. 1429
FRIEDMAN, J. 1392
FRIEND, D.J.C. 5619
FRIIS-NIELSEN, B. 245
FRIMMEL, G. 2798, 4121
FRISSEL, M.J. 6511
FRITSCHEN, L.J. 2799, 3048,
5620
FRITTS, R., Jr. 4122
FRITTON, D.D. 3875
FRÖHLICH, H. 5358
FRÖLICH, W.G. 2800
FROMMHOLD, I. 246, 247, 1393,
2801, 2802
FROTA, J.N.E. 4123, 4124
FRYER, J.T. 1881
FUCHIGAMI, L.H. 4657
FUCHS, M. 2274, 2803, 3165,
3784, 5799
FUEHRING, H.D. 248
FUJIMOTO, K. 5621
FUJIOKA, R.S. 4125
FUKAI, S. 5898
FUKUDA, M. 2213, 3444
FUKUEI, K. 2139
FUKUYA, A. 4233, 4234
FULLER, M.M. 270
FUNADA, S. 4233, 4234, 4235
FURCH, B. 4126
FURUTA, T. 1394

G

GABR, A.I. 208, 249
GABRIELIAN, J.Y. 3740
GABRIELS, D. 2703, 5843
GABRIËLS, R. 5413
GABRIELYAN, A.G. 998
GAFF, D.F. 310, 1395, 1396,
2421, 2422, 2423, 2804, 5622,
5623, 5687, 5688, 6497
GAFFNEY, J.J. 4127
GAJRI, P.R. 692, 4784
GALATIS, B. 2805
GALBRAITH, K.A. 114, 115
GALE, J. 250, 251, 252,
253, 676, 1397, 1644
GAŁECKA, B. 4128
GALEEV, N.A. 4583
GALES, K. 4331, 5402
GALIEV, N.A. 2378, 2379, 4129
GALLACHER, A. 826, 2268, 4130
GALLAGHER, J.N. 1398, 3805,
5624, 5625
GALLAHER, R.N. 2806, 2807,
3196
GALLETLY, W.S. 4937
GALLI, E. 6473
GALLI, M.C. 4131
GALLOWAY, R.A. 5789
GALSON, E. 2405, 3633
GALSTON, A.W. 2154, 2171, 2230,
3315, 4794
GALSTON, E. 6360
GALUN, M. 3446, 6342
GAMEZ, H. 6259
GAMPER, L. 2448
GAMZIKOVA, O.I. 5626
GANCHARYK, M.M. 2808
GANCHEV, G.T. 4132, 4133
GANDAR, P.W. 1399, 1400, 1401,
1402
GANEA, V. 6300
GANGADHARA, M. 254, 361, 1089,
1591, 1592, 4134, 4135, 4173,
4304, 4305
GANZLIN, G. 4136
GARAFEEV, A.G. 2750
GARANOVICH, I.M. 1115, 1403,
5627
GARBER, M.P. 2809, 4477
GARCIA, M. 1404
GARCÍA NOVO, F. 1838, 2058
GARCIA-RINCON, G.E. 1405
GARDI, I. 2305
GARDNER, B.R. 1406, 3360
GARDNER, H.R. 224
GARDNER, W.R. 154, 2810, 2902,
2903, 3967
GARLAND, J.A. 4137
GARLASCHI, F.M. 2548
GARNSEY, S.M. 1407, 2470, 3668
6466

MAMARIL, C.P. 5346
MAMULASHVILI, G.G. 515, 1797
MANALO, J.R. 2114
MANAM, R. 3100
MANANKOV, M.K. 516
MANAWI, M. 3398
MANCHANDA, H.R. 3101
MANCHESTER, J. 6123, 6124
MANDAL, R.K. 3225
MANG, H.A. 2667
MANJREKAR, S.P. 407
MANKIN, J.B. 866, 2296, 3081
MANN, H.S. 5985
MANN, J.D. 3040, 4464
MANN, L.K. 6005
MANNETJE, L.'t. 5192
MANNING, C.E. 3102, 3150
MANNING, D.M. 1391
MANNINGER, K. 4545
MANNINGER, S. 4545
MANOGARAN, C. 517
MANOHAR, M.S. 3103, 3104
MANOHAR, S.S. 1714, 1955
MANOLAKIS, E. 3105
MANSELL, R.S. 211, 4844
MANSFIELD, T.A. 1366, 1798,
1799, 1800, 1801, 1944, 2766,
3106, 3575, 3978, 4546, 5190,
5986, 5987, 6244, 6478, 6479
MANSOUR, N.S. 2543
MANTELL, A. 5988
MANUIL'SKIJ V.D. 615
MANYATIN, Yu.K. 4547
MAOTANI, T. 1802, 1803, 1804,
1805, 2890, 3086, 3107, 3108,
3109, 3110, 5989,
MAPP, H.P., Jr. 518, 1806
MARAIS, J.N. 519
MARANGONI, B. 4819, 5115
MARBACH, I. 520, 6543
MARC, J. 1807
MARCELLE, R. 4673, 6508
MARCELLOS, H. 3111
MARCESSE, J. 1884
MARCHAND, P.J. 4548
MARCO, G., di 1808
MARCZYŃSKI, S. 4549, 5990
MARETZKI, A. 3112, 4607
MARFINA, K.G. 750, 2142,
3386
MARGARIS, N.S. 3113
MARIANI, G. 4550
MARIE, R. 1071
MARIEN, J.N. 1071
MARINCHIK, A.F. 521
MARINO, M.A. 3692
MARK, A.F. 522, 5991
MARKGRAF, G. 6178
MARKHART, A.H. III. 4551, 5595,
5992
MARKOVA, M.N. 4373
MARKOWSKI, A. 1809
MARKS, C.F. 2348

MARRÉ, E. 3114
MARREWIJK, N.P.A. 4552
MARSCHNER, H. 3154, 6260
MARSH, A.W. 1394, 1489, 1490,
1810
MARSHAKOVA, M.I. 523, 2947
MARSHALL, D.R. 2591
MARSHALL, P.E. 3115
MARSHO, T.V. 4553
MARTENS, J. 4554
MÄRTIN, B. 4269
MARTIN, C.E. 1811
MARTIN, C.K. 5993
MARTIN, E.S. 524, 3511, 3512,
3513, 5021, 5994, 6410
MARTIN, F.G. 1771, 3065
MARTIN, G. 5995
MARTIN, G.C. 1198, 3430, 3898,
5724
MARTIN, J.K. 3116
MARTIN, J.P. 5601
MARTIN, P.E. 1281, 1282
MARTIN, P.J. 474
MARTIN, R.E. 525
MARTIN, R.J. 526
MARTIN, T.J. 527
MARTIN, W.H. 1812
MARTINEZ, T. 4050
MARTINEZ-CARRASCO, R. 5996
MARTYNIAK, B. 5894
MARX, A. 3117
MARYNICK, D.S. 528
MARYNICK, M.C. 528
MARZOLA, D.L. 5997
MASAMOTO, K. 4555
MASAROVIČOVÁ, E. 1315
MASARYKOVÁ, V. 1840
MASAYOSHI, F. 529
MASON, G.F. 530
MASON, W.K. 3869
MASSIMINO, D. 3717, 5273
MASSIMINO, J. 5273
MASTERSON, C.L. 4556
MASUDA, Y. 4557, 4558, 6596
MATAR, A.E. 3118
MATĚJÍKOVÁ, O. 4559, 5299
MATHERS, A.C. 3119
MATHRE, D.E. 3120
MATHUR, D.D. 1813
MATI, R. 5956
MATIČ, M. 3286
MATORIN, D.N. 3201
MATSCHKE, C. 2157
MATSUDA, K. 2092, 2874, 2875,
4845
MATSUDA, M. 6283
MATSUDA, T. 1656
MATSUMOTO, K. 4366
MATSUOKA, Y. 5998
MATSUYAMA, S. 5760
MATSYUSHEĬSKAYA, V.P. 2808
MATTEUCCI, S.D. 1814
MATTHEWS, J.M., Jr. 5999

MELKÖ. E. 4779
MELVILLE. M.D. 141. 3944
MELVILLE. R. 1836
MENGE. J.A. 4825
MENGEL. K. 6013
MENOUX-BOYER. Y. 6014
MENYAÏLO. L.N. 4579
MEON. S. 6015
MEREDITH. P. 1837
MÈRIAUX. S. 544 2579
MERINO. J. 1838 2058
MERRIGAN. I.F. 6544
MERRILL. S.D. 4580. 6016
MERT. H.H. 6017
MERTIA. H.S. 2523
MERYA. G.E. 3178
MESICEK, J. 348
MESSEM, A.B. 545, 2720
METCALF, C. 149
METCALF, C.L. 4266
METHY, M. 195, 4051
MÉTRAUX, J.-P. 5055
METTAUER, H. 5437
METZNER, H. 3684
MEUSEL, M. 5497
MEXAL, J. 546
MEXAL, J.G. 3343
MEYER, A. 5910
MEYER, C.P. 1839
MEYER, H.R. 4335
MEYER, W.S. 6018, 6019, 6249
MEZENTSEVA, V.T. 547
MICHAEL, G. 1434
MICHAELIS, G. 2920
MICHALCZYK, K.-W. 6020
MICHALOV, J. 1840
MICHAUX-FERRIÈRE, N. 4581
MICHEL, B.E. 209, 548, 1841,
2739, 3141, 3142, 3226, 4582,
6021
MIDDLETON, J.E. 3306, 6022
MIEKELEY, W. 5245
MIFTAKHUTDINOVA, F.G. 32, 549,
1842, 3725, 3726, 3727, 4583
MIGUNOVA, E.S. 1843, 1844
MIGUSHOVA. É.F. 3755
MIKHAÏLOV, M.V. 550
MIKHAÏLOVA, A.V. 3143, 6023
MIKHAÏLOVA, L.M. 3832
MIKHAÏLOVA, S.P. 3044
MIKHALEV, S.S. 3719
MIKHALIK, E. 4403
MIKOLAJCZYK, S.D. 5567
MIKUL'SKAYA, S.A. 491, 1598,
2808, 6503
MILBORROW, B.V. 4584, 4585,
6024
MILBURN, J.A. 974, 3144,
3145, 6025
MILBURN, T.R. 4355
MILELLA, A. 4586
MILES, D.L. 6
MILES, G.E. 3146

MILES, W.G. 3147
MILFORD, G.F.J. 487, 551, 552,
553, 3148, 4042
MILICÅ, C.I. 554. 1963
MILL. R.B. 2169
MILLAR. B.D. 555. 1299. 1300.
1845. 2705
MILLER. D.A. 6524
MILLER. D.E. 1846. 1847. 3149
MILLER. D.G. 3102. 3150
MILLER, D.M. 4587, 5086, 5129,
5219, 5220, 6585
MILLER, E.E. 2902, 2903
MILLER, E.L. 1848
MILLER, F.R. 2902, 2903
MILLER, G.E. 90, 1849
MILLER, J.H. 2969, 2970
MILLER, J.J. 4270
MILLER, L.D. 3584
MILLER, L.N. 1169, 2605, 5388
MILLER, M.G. 1822
MILLER, M.H. 3712
MILLER, N.A. 4588
MILLER, P.C. 198, 199, 556,
679, 845, 1850, 1851, 4589,
4767, 6026, 6027, 6108, 6415
MILLER, P.R. 1852
MILLER, R.D. 3151
MILLER, R.F. 1142
MILLER, R.H. 5182
MILLER, R.J. 1458
MILLER, R.L. 525
MILLER, R.W. 6028, 6361
MILLER, T.D. 2483, 2484
MILLET, B. 4590
MILLIKAN, D.F. 1231
MILLINGTON, A.J. 3650
MILLKOVÁ, J. 769
MILLS, J.D. 6029
MILLS, J.T. 6030
MILLS, P.D. 3934
MILNE, R. 3987, 6031
MILTHORPE, F.L. 139, 557, 1010,
1180, 1853, 1868, 3567, 4208,
5082, 5187, 6032, 6033, 6460
MINCHIN, F.R. 558, 1854, 3574,
4591
MINCHIN, P.E.H. 4751
MINGEAU, M. 1323, 1874
MINKOV, I. 3152
MINKOV, I.N. 1676, 5837
MINOR, H.C. 5169
MINSHALL, W.H. 559, 560, 561
MIRHADI, M.J. 6034
MIROCHA, C.J. 1223
MIROSHNICHENKO, Yu.M. 6035
MIRZAEV, A. 4592
MIRŽINSKI-STEFANOVIĆ, L. 5011
MISAGHI, I.J. 4593, 6036
MISHKIND, M. 6037
MISHNEV, V.G. 1855
MISHRA, D. 5523, 5524
MISRA, B. 5933

RAKHIMOV, G.T.   2055, 2056
RAKHIMOVA, T.   4810
RAKHMANINA, A.T.   705, 4811,
4812
RAKHMANINA, K.P.   706
RAKHTEENKO, L.I.   1599
RAKITIN, L.Yu.   707
RAKITIN, V.Yu.   707
RAKITINA, Z.G.   4813
RAKOV, K.   2057
RAKOVAN, J.   5483, 5484
RAMACHANDRAM, M.   3326, 4816,
6220
RAMACHANDRA PRASAD, T.V.   1716,
1717
RAMA DAS, V.S.   6218   see also
DAS, V.S.R.
RAMAKRISHNA, T.M.   4814
RAMAKRISHNA, Y.S.   804, 3463,
4968
RAMAKRISHNAN, P.S.   4815
RAMA KRISHNAYYA, G.   6219
RAMA MOHAN RAO, M.S.   4816,
6220, 6221
RAMANATH, B.   6516
RAMATI, A.   6222
RAMAYYA, N.   3325
RAMAZANOVA, L.Kh.   4817
RAMESH BABU, V.   4818
RAMEY, H.H., Jr.   6384
RAMIG, R.E.   727, 4855, 6202
RAMINA, A.   4156
RAMIREZ, J.M.   3769
RAMIREZ DIAZ, L.   2058
RAMOS, C.   5452, 6223
RAMOS, D.E.   4819, 5115
RAND, R.H.   2639, 2667, 3323
RANDAZZO, G.   2526
RANDHAWA, G.S.   4538
RANEY, R.J.   3365
RANGA RAO, V.   4816, 6220
RANI, S.   2868
RAO, A.S.   4820
RAO, B.U.C.   5130
RAO, D.V.M.   6224
RAO, I.M.   1274, 1275, 2052,
2053, 2054, 2690, 6217
RAO, J.S.   2325
RAO, J.V.S.   6218
RAO, M.J.   2059
RAO, M.S.R.M.   3326
RAO, P.N.   3322, 3324
RAO, S.R.S.   3325
RAO, T.B.   4135
RAO, T.V.   3097
RAO, V.R.   3326
RAPER, C.D., Jr.   708, 5147
RAPP, M.   4821, 5747
RÄSÄNEN, P.K.   5689
RASCHKE, K.   709, 710, 711,
712, 2060, 2061, 2062, 2063,
2064, 2713, 2714, 3327, 3328,
3329, 4035, 4083, 4755, 4822,
4823, 4882, 5122, 5123, 5574,
6225, 6226   see also

RASHKE, K.   2065
RASMUSSEN, H.P.   1623
RASMUSSEN, O.S.   2066
RASMUSSEN, V.P.   1499, 2899,
4212, 4824, 5801, 6227
RASMUSSON, D.C.   316, 968
RATHAIAH, Y.   2067, 3330, 3331
RATHNAN, C.K.M.   3332
RATNAM, B.P.   3333
RATNASOORIYA, G.B.   3182
RATNASURIYA, G.B.   4636
RATNAYAKE, M.   4825
RATUSHNYAK, Yu.M.   228, 403,
425, 4826
RAUNER, Yu.L.   6228
RAUSCHER, H.M.   4827
RAUSCHKOLB, R.S.   3247
RAUZI, F.   2886
RAVELLI, F.   4490, 4828
RAVELOMANANA, D.   1071
RAVEN, J.A.   713, 2068, 2252,
3334
RAVINDRANATH, E.   3235
RAWITZ, E.   6229
RAWLINS, S.L.   346, 714, 2069,
6016
RAWSON, H.M.   715, 2070, 2071,
2454, 3335, 3336, 4829, 5105,
5788, 5968, 6230
RAY, L.L.   2049
RAY, S.   3037
RAY, T.B.   6231
RAYCHAUDHURI, S.P.   4539
RAZI, B.A.   4814
READ, D.J.   2811
REBELLA, C.   3560
REBOUCAS FERREIRA, L.G.   4786
REDDELL, D.L.   350
REDDY, A.R.   6232
REDDY, D.S.   5933
REDDY, P.K.R.   6233
REDMANN, R.E.   2072, 3349, 4561,
4562, 4830, 4831
REED, K.L.   1688
REED, M.L.   1097
REED, R.H.   6234, 6235
REESE, R.L.   1699, 2073, 3004,
3337, 3338
REETZ, H.F., Jr.   6236
REEVES, H.E.   716, 717, 718,
719
REEVES, T.G.   2074
REFFYE, P., de   2075
REGEHR, D.L.   720
REGER, B.J.   4037
REGIER, C.   2214
REGINATO, R.J.   360, 721,
1585, 1586, 1587, 1588, 1607,
1608, 1675, 2076, 2077, 2416,
2938, 2950, 3339, 4061, 4062,
4298
REHM, G.W.   3340
REICHEL, E.   1069
REICHMAN, G.A.   3794, 4111

STONE, J.F.   518, 716, 717, 790, 843, 2215, 2216, 2217, 2647, 4566, 5558, 6413
STONE, L.R.   844, 1642, 1816, 1817, 2285, 2966, 3365, 6414
STONER, W.A.   845, 1851, 4589, 6027, 6415
STOREY, J.B.   4318
STOREY, R.   846, 3664, 5031, 5032, 5206, 5207, 5208, 6416
STOSZEK, K.J.   5007
STOUT, B.B.   3659
STOUT, D.G.   3524, 3525, 3912, 4369, 5033, 5034, 5035, 5036, 5679, 6349, 6417
STOUT, R.   3932
STOYANOV, Zh.V.   847, 848, 4642
STOYNOVA, E.   4644
STRAIN, B.R.   849, 3068, 4084, 6255, 6442
STRATEENER, G.   789, 850
STREBEL, O.   851, 4837, 5037, 5038, 5120
STREBEYKO, P.   3526
STREHLOW, H.   87
STREICH, J.   2716
STREL'TSOVA, V.S.   4008
STRNADOVÁ, H.   2185
STROGONOV, B.P.   5298
STROHMAN, R.   1394
STROMBERG, L.K.   5281, 5282, 6528
STROOSNIJDER, L.   3431
STRUCHTEMEYER, R.A.   2235
STUART, D.A.   2286, 3527, 5039
STUART, K.L.   140, 852
STUART, M.   5868
STUART, M.L.   3443
STUBBENDIECK, J.   194
STUBBLEBINE, W.H.   5911
STUBBLEFIELD, S.   5040
STUCKEY, R.E.   527
STUMM, G.   3528
STUMPE, G.   5657
STUMPF, D.K.   6399
STUPISHINA, E.A.   423, 853, 3913
STURGES, D.L.   3529, 6418
STUSHNOFF, C.   1178
STUTTE, C.A.   1248, 5041, 5460, 6419, 6541
STYER, R.C.   2287
SUAREZ, J.J.   3530
SUBBA RAO, K.   5130
SUCOFF, E.   854, 855, 1350, 1553
SUD'INA, E.G.   3531
SUDNITSYNA, T.N.   6420
SUDO, K.   3532
SUGAHARA, K.   4424
SUGIMOTO, H.   5056
SUGIMOTO, K.   856, 857, 2288, 3533, 3534

SUGIYAMA, T.   6421
SUHAIL, B.A.   5402
SUKUMARAN, N.P.   858
SULEYMANOV, I.G.   82, 422, 423, 425, 797, 859, 860, 2992
SULLIVAN, C.Y.   6422
SULLIVAN, D.M.   1570
SULLIVAN, T.E.   2882
SULLIVAN, T.P.   861
SULYALINA, A.V.   5042
SUMAYAO, C.R.   3535, 6423
SUMMERFIELD, R.J.   1854, 3574, 4591
SUMNER, D.R.   5776
SUNDQVIST, C.   3788, 5043, 6424
SUNG, F.J.M.   6425
SUOMELA, H.   207
SURÁNYI, D.   862
SURMA, M.   1051
SUSSEX, I.   2289
SUTCLIFFE, J.   3536 see also
SUTCLIFFE, J.F.   2290, 2291, 2664, 3537
SUTER, E.   1551, 5044
SÚTOR, J.   3538
SUTTON, B.G.   863
SUTTON, J.C.   5045
SUZDAL'TSEVA, V.A.   864
SUZUKI, K.   5046
SUZUKI, M.   2906
SUZUKI, T.   4602
ŠVACHULA, V.   2292, 3539, 3540, 3541, 6426
ŠVACHULOVÁ, J.   3540, 6426
SVEINBJÖRNSSON, B.   4589, 6109
SVENSSON, B.   3542
SVENSSON, J.G.P.   5448
SVENSSON, S.-B.   6498
SVESHNIKOVA, V.M.   865, 2293, 2294, 5047, 5048, 5049, 6427, 6428
SVIHRA, J.   3543, 3544
SVOBODOVÁ, J.   3545
SWAMY, P.M.   1275
SWANK, W.T.   866, 2296
SWANSON, C.A.   2295
SWANTON, C.J.   5045
SWARD, R.J.   1715
SWARUP, V.   778
SWEENEY, B.M.   6429
SWIFT, J.G.   1186
SWIFT, L.W., Jr.   866, 2296
SWINGLE, H.D.   5999
SYBER, A.Yu.   3546
SYDENHAM, P.H.   867
SYLVIA, D.M.   5050
SYME, J.R.   2778
SYNNATSCHKE, G.   2860
SYRATT, W.J.   1462
SYVERTSEN, J.P.   868, 5051, 5453, 6430
SZABÓ, M.   2297

W

WABER, J. 935, 2393, 5144
WACHTER, A. 936
WACQUANT, J.-P- 734, 6523
WADDINGTON, D.V. 940
WADDLE, B.A. 6384
WADE, G.C. 1673
WADSWORTH, R.M. 1349
WAGGONER, P.E. 639, 2394,
2395, 2396
WAGNER, E. 5145
WAHAB, A. 2397
WAHUA, T.A.T. 6524
WAIN, R.L. 1737, 6525
WAINES, J.G. 5686
WAINWRIGHT, S.J. 996, 1535,
2489
WAISEL, Y. 978, 3446, 6222,
6342
WAKEFIELD, R.C. 4655
WAL, A.F., van der see
Van der WAL, A.F.
WALCOTT, J. 574
WALCOTT, J.J. 3632
WALKER, D.A. 496, 2399
WALKER, G.K. 6526
WALKER, J.N. 2425, 2426
WALKER, N.A. 2400, 2401
WALKER, R.B. 4488, 5929
WALKER, R.H. 5782
WALKER, R.R. 6527
WALKER, T.W. 1509
WALKER, W.R. 3471
WALL, B.H. 2948
WALLA, J.A. 6180
WALLACE, A. 46, 1684, 1685,
1686, 1975, 2402, 3244, 3733,
4415
WALLACE, D.H. 3219, 3249,
3250
WALLACE, H.A.H. 6030, 6370
WALLACE, H.R. 6015
WALLACE, J.S. 1096, 5625
WALLACE, W.D. 2345
WALLACH, D. 2403, 2404
WALLEN, V.R. 3768
WALLENDER, W.W. 6528
WALLIHAN, E.F. 3244, 5146
WALSBY, A.E. 2847
WALTER, A. 3709
WALTON, D. 2405
WALTON, D.C. 322, 2406, 3633,
6360
WALTON, P.D. 937, 2309
WAMPLE, R.L. 938, 2407, 2408
WANDERLINGH, F. 4155
WANN, M. 5147
WARD, C.R. 5417
WARD, M.E. 4108
WARDER, F.G. 1452, 2617
WARDLAW, I.F. 2409, 2410,
3485, 5564

WARDLE, K. 6529, 6530
WAPDOWSKI, W.F. 1453
WARE, G. 5008
WAREING, P.F. 6531
WARGO, P.M. 939
WARING, R.H. 737, 1688, 2411,
2412, 2743, 5148, 6532, 6533
WARMBRODT, R.D. 5149
WARRELL, L.A. 6006
WARREN, R.C. 2413
WARRICK, A.W. 5316, 5949, 5950,
6534
WARRINGTON, I. 2414
WARRINGTON, I.J. 3634, 5150
WARRIT, B. 474, 1732
WÄSTERLUND, I. 5775
WATANABE, A. 4303
WATANABE, I. 6343
WATANABE, K. 2511
WATANABE, S. 4299, 5210, 5251,
6583
WATKIN, B.R. 1540
WATSCHKE, T.L. 940
WATSON, C.A. 272, 273
WATSON, E.R. 2415
WATSON, K.K. 941, 2416
WATSON, L. 5151
WATT, L.A. 6535
WATTERS, G. 3883
WATTS, D.G. 4989
WATTS, W.R. 474, 2417, 3635,
5152
WEARING, A.H. 6536
WEATHERLEY, P.E. 942, 2418,
2757, 5569, 6537
WEATHERLY, A.B. 6538
WEATHERSPOON, D.M. 5153
WEAVER, J.B., Jr. 2059
WEAVER, R.E.C. 6004
WEAVER, T. 3943
WEBB, J.R. 5562
WEBB, W. 5154
WEBER, J. 5912
WEBER, M. 544
WEBSTER, B.D. 3636, 4611, 6055
WEBSTER, G. 5478
WEBSTER, G.R. 1511
WEBSTER, R. 3637
WEEKS, D.L. 6292
WEETE, J.D. 4487, 5155
WEGMANN, K. 4624, 5156, 5157
WEIBEL, D.E. 4666, 4667
WEIGER, C. 3861
WEIHE, K., von 6539, 6540
WEILAND, R.T. 5041, 6419, 6541
WEINBERG, R. 943
WEINBERGER, P. 5158, 5159
WEINMANN, R. 944
WEISBROD, M. 789
WEISER, C.J. 128, 4229
WEISS, A. 5345
WEISSER, P.J. 3161
WEISSHAUPT, F. 3638

PLANT INDEX

(Cumulative index to volumes 1 - 5)

This index contains plant genera and types interesting as experimental material for physiological, ecological and agricultural studies. The Latin plant names are the main items which present the reference number. English names of the most common plants are cross-indexed.

A

Abelmoschus 3422

Abies 498, 499, 697, 977, 1037, 1146, 1250, 1403, 1510, 1530, 1755, 1852, 2013, 2128, 2317, 2411, 2542, 2635, 2675, 2676, 2854, 3702, 3737, 3744, 4010, 4094, 4182, 4434, 4548, 4565, 4955, 5227, 5237, 5762, 5768, 5943, 6477, 6494

Abronia 3992, 3993, 3994, 5499

Abrus 399, 2138, 5804, 6321

Abutilon 1966, 2008, 2432, 3855, 4729, 4730, 4731, 4752, 4753, 5875, 6155, 6218

Acacia 118, 123, 908, 1134, 1169, 2037, 2038, 2362, 2603, 2767, 3414, 3437, 3933, 4349, 4629, 5054, 5358, 6568

Acalypha 2691

Acanthus 5256, 6321

Acer 129, 149, 160, 161, 204, 355, 356, 568, 674, 824, 1097, 1235, 1269, 1345, 1362, 1363, 1389, 1403, 1577, 1578, 1755, 1812, 1947, 2019, 2020, 2094, 2128, 2193, 2254, 2317, 2347, 2509, 2586, 2588, 2603, 2686, 2698, 2734, 2735, 2737, 2761, 2933, 2934, 2935, 3115, 3262, 3356, 3564, 3582, 3585, 3737, 4064, 4182, 4266, 4297, 4434, 4450, 4462, 4463, 4499, 4540, 4543, 4627, 4648, 4650, 4742, 4768, 5106, 5119, 5174, 5239, 5297, 5393, 5546, 5547, 5548, 5549, 5555, 5579, 5727, 5746, 5765, 5766, 5767, 5779

Achilea 1005, 1938, 2685, 3715, 3943, 5499, 5878

Achras 1185

Aconitum 3728

Acorus 3772, 4432, 4433

Adenocarpus 2138

Adoxa 4504

Aegiceras 2962

Aegle 4629

Aegopodium 568, 3728, 4732

Aellenia 2293, 4810,

Aeonium 6286

Aeschynomene 1277, 1705, 5070, 5804, 6321

Aesculus 5212, 5555, 6494

Agathis 2377, 2984

Agave 259, 597, 1337, 1506, 1916, 2814, 3204, 4663, 5435, 5544, 5545, 5636, 5953, 6103, 6452, 6455

Agropyron 79, 171, 532, 1005, 1118, 1387, 1557, 1826, 2072, 2567, 2660, 2770, 2954, 3128, 3397, 3715, 3814, 3821, 3943, 4236, 4293, 4348, 4390, 4391, 4561, 4562, 4616, 4830, 4831, 5049, 5077, 5127, 5184, 5331, 5499, 5707, 5769, 5778, 6056, 6246, 6268, 6351, 6352, 6353, 6476, 6523, 6560

Agrostis 389, 390, 468, 996, 1457, 2489, 3379, 3704, 4240, 4456, 4655, 5077, 5151, 5182, 5202, 5599, 6071

Ailanthus 1711, 3115

Alagonium 2868, 5804

Albizzia 2171, 3884, 4367, 4381, 5061

Alchemilla 1419

alder see *Alnus*

alfalfa see *Medicago*

Algae
    Alaria 3863
    Amphidinium 5731
    Anabaena 2847, 4955
    Anacystis 289, 798
    Aphanothece 4587, 5086, 5129,
    5219, 5220, 6585
    Ascophyllum 4913, 5371, 5502,
    5816
    Bostrychia 3983
    Bryopsis 5816, 5845
    Caulerpa 5205, 5845
    Ceramium 5816
    Chaetomorpha 4381, 5816
    Chara 236, 434, 976, 1774, 1784,
    3626, 4413, 5242, 5816, 6602
    Chlamydomonas 713, 1888, 3173,
    4955
    Chlorella 132, 665, 713, 902,
    1514, 1562, 1630, 1999, 2962,
    3680, 4145, 4598, 4747, 5035,
    5063, 5663, 5664, 6325
    Chondrus 5887
    Cladophora 3983
    Codium 60, 72, 1098, 2869, 3937,
    4381, 5816
    Corallina 5845
    Cyanophora 3401
    Cyclotella 1766, 1767, 1768,
    5816
    Dunaliella 700, 1076, 1150,
    3830, 3846, 3848, 3861, 4113,
    4154, 4358, 4381, 4624, 5157,
    5298, 5376, 5646, 5647, 5789,
    5816
    Enteromorpha 2962, 4874, 6234,
    6235
    Eremosphaera 5639
    Euglena 1149, 4955, 5504
    Fucus 4913, 5502, 5787, 5861,
    5835, 6081, 6207
    Gracilaria 3983
    Halicystis 1449, 3869, 5816
    Halimeda 5845
    Hydrodictyon 5816
    Hypnea 5845
    Laminaria 5502, 5787
    Microcystis 4904
    Nannochloris 200
    Nitella 433, 976, 1562, 1600,
    1784, 2318, 2764, 2765, 3215,
    5020, 5816
    Nostoc 2958, 3358
    Ochromonas 408, 1562, 4381,
    4382, 5816
    Oedogonium 2962
    Pelvetia 4913, 5502
    Phaeodactylum 6309

Algae (continued)
    Platymonas 429, 430, 1527, 1678,
    2996, 2997, 4381, 5816
    Pleurococcus 909, 1420
    Porphyra 5887
    Porphyridium 2827, 4598
    Rhodymenia 5845, 5887
    Sargassum 1629
    Scenedesmus 713, 5816
    Spirogyra 5639
    Stichococcus 3851, 3852, 3853,
    3854
    Stigeoclonium 3852
    Synechococcus 200, 4904
    Thalassiosira 5731
    Trentepohlia 1420
    Ulothrix 3852
    Ulva 1629, 6207
    Valonia 1562, 2671, 2869, 2891,
    3509, 3680, 3681, 3709, 3793,
    4381, 5816
    Vaucheria 801, 6348

Alhagi 1705

Alisma 4504, 4616, 5975

Alliaria 4768

Allium 250, 259, 806, 1152, 1341, 1478,
1514, 1520, 1681, 1682, 1842, 1869,
1893, 1998, 2032, 2083, 2094, 2110,
2146, 2167, 2179, 2198, 2287, 2294,
2317, 2337, 2430, 2474, 2724, 2727,
2772, 3183, 3224, 3228, 3229, 3230,
3231, 3232, 3233, 3346, 3357, 3367,
3406, 3673, 3676, 3685, 3815, 3916,
3929, 4149, 4153, 4173, 4174, 4469,
4583, 4605, 4625, 4704, 4707, 4709,
4738, 4770, 4771, 4811, 4876, 4878,
4879, 4907, 4908, 4980, 5020, 5045,
5048, 5088, 5231, 5358, 5501, 5919,
5967, 5978, 6133, 6225, 6265, 6292,
6311, 6336, 6410, 6452, 6455, 6494,
6600

almond see *Amygdalus*

Alnus 265, 1169, 1235, 2603, 2605,
2761, 2787, 4367, 4499, 4732, 5388,
5721, 5869, 5870, 5871, 6477, 6598

Alocasia 65, 74

Aloë 464, 1630, 1726, 2430, 2962,
4359, 4449, 4450, 5363, 5435

Alopecurus 18, 1005, 2835, 3379, 3380,
4006, 4220, 4456, 4881, 6513

Alpinia 1591

Alstonia 1185

Asperugo   1862

Asperula   203, 1342, 1343, 1344,
4064, 4768

Aspidosperma   2961

Aster   675, 898, 2608, 3379, 3704,
4811, 5332, 6172, 6218

Astilbe   3164

Astragalus   4808, 5445, 5769, 5979,
6396

Astrebla   133, 1310, 3153, 3298,
5103, 5622

Atriplex   65, 74, 75, 250, 675, 846,
902, 1078, 1097, 1134, 1142, 1451,
1562, 1684, 1791, 1934, 2194, 2424,
2476, 2614, 2881, 3284, 3336, 3437,
3440, 3442, 3593, 3664, 3704, 3855,
3992, 3993, 3994, 4092, 4101, 4665,
4818, 5208, 5318, 5329, 5331, 5332,
5475, 5499, 5659, 5686, 5769, 5882,
5952, 6009, 6049, 6198, 6199, 6231,
6297, 6395, 6416, 6428

Aucuba   4182, 5809

Avena   37, 97, 98, 138, 166, 175,
176, 212, 213, 267, 287, 491, 925,
1032, 1067, 1068, 1079, 1151, 1245,
1289, 1302, 1429, 1498, 1514, 1528,
1556, 1624, 1625, 1661, 1693, 1694,
1773, 1837, 1887, 1888, 1895, 1896,
1897, 1934, 2011, 2083, 2089, 2126,
2127, 2143, 2148, 2196, 2225, 2290,
2301, 2317, 2360, 2415, 2464, 2596,
2624, 2625, 2808, 2879, 2930, 2976,
3014, 3063, 3067, 3098, 3179, 3315,
3372, 3388, 3389, 3425, 3437, 3496,
3592, 3698, 3699, 3747, 3749, 3789,
3802, 3875, 3923, 3926, 4034, 4035,
4071, 4321, 4343, 4381, 4392, 4416,
4558, 4843, 4893, 4906, 4977, 5028,
5090, 5270, 5366, 5372, 5427, 5504,
5659, 5758, 5769, 5781, 5919, 5927,
5967, 6061, 6066, 6178, 6225, 6321,
6335, 6370, 6372, 6373, 6408, 6448,
6449, 6450, 6452, 6453, 6455, 6494,
6503, 6574

Avenastrum   4328

Avicennia   556, 675, 2962, 4336, 6308

Azadirachta   2951

Azalea   3971, 5153, 5413

B

Bacteria
   Bacillus   3846, 4381, 5601
   Chromatium   135, 4425
   Clostridium   4381, 5601
   Corynebacterium   2249, 5486
   Erwinia   1992, 2249, 3399
   Escherichia   6411
   Halobacterium   1046, 3848, 4941,
   5601, 5850, 6462, 6591
   Lactobacillus   3846
   Pseudomonas   1749, 1859, 2158,
   2828, 5601, 5675
   Rhizobium   2455, 3579, 5702
   Rhodopseudomonas   3931
   Staphylococcus   3846
   Streptococcus   6411
   Xanthomonas   2158, 2249, 2428,
   5675, 6405

Baldellia   1882

Bambusa   1999

banana see *Musa*

Banksia   3437

barley see *Hordeum*

Barringtonia   5571

Basella   6187

Batis   5900, 6396

bean see *Phaseolus* or *Vicia*

beech see *Fagus*

Begonia   1289, 1514, 2471, 5382, 5572

Bellis   4456

Berberis   1021, 4611, 4655

Bergenia   3164

bermudagrass see *Cynodon*

Beta   38, 63, 277, 280, 307, 453, 487,
521, 551, 552, 553, 585, 592, 611, 614,
620, 666, 800, 810, 815, 851, 899, 915,
1074, 1097, 1117, 1121, 1134, 1205,
1206, 1269, 1311, 1334, 1371, 1412,
1414, 1451, 1469, 1486, 1514, 1520,
1536, 1552, 1755, 1785, 1846, 1867,
1869, 1934, 1960, 2015, 2022, 2067,
2083, 2097, 2110, 2128, 2133, 2156,
2157, 2161, 2162, 2276, 2292, 2317,
2382, 2383, 2476, 2538, 2613, 2626,
2693, 2706, 2707, 2720, 2747, 2758,

Castanea 4367, 5050, 5809

Casuarina 632, 1030, 5195, 6568

Cattleya 2836

Ceanothus 679, 3012, 3064, 3871, 4442, 6026

cedar see *Tamarix*

Cedrus 1037, 1635, 3599, 3744

Celastrus 646

Celosia 2471

Celtis 4367

Cenchrus 133, 134, 647, 2315, 2447, 4324, 5192, 6158, 6159, 6195, 6359, 6563

Centaurea 4779

Cerastium 4456

Cerasus 562, 698, 916, 1021, 1913, 2462, 2540, 5139, 5284, 6107

Ceratonia 1208, 2362, 4016

Ceratophyllum 5962

Cerbera 1185, 5098

Cercocarpus 2006

Cereus 5087

Chaerophyllum 4732

Chamaecyparis 5297, 5943, 6604

Chamaenerion 3728, 6507

Cheiranthus 1893, 6556

Chenopodium 1625, 1966, 2317, 3539, 3704, 3855, 4752, 4753, 5145, 5212, 5499, 5863, 6395

cherry see *Cerasus*

Chloranthus 5902

Chloris 846, 3611, 4324, 5126, 5947, 6395

Chlorophytum 5442

Chrysanthellum 1576, 1938

Chrysanthemum 309, 825, 1117, 1514, 1780, 1782, 2108, 2471, 2817, 3399, 3747, 4407, 4536, 4549, 5769, 5994, 6060, 6251, 6252

Chrysosplenium 3164

Chrysothamnus 1142

Cicer 1485, 2192, 2872, 4337, 4430, 4803, 4874, 4876, 4894, 5135, 5804, 6321, 6367

Cichorium 3922, 4732, 5769, 6265

Cinnamomum 4367

Cirsium 6574

Cistus 1696, 1838, 2375, 3113

Citrulus 478, 805, 981, 2430, 2552, 3437, 4513, 4818, 4971, 6322

Citrus 145, 161, 305, 323, 326, 443, 458, 493, 608, 614, 810, 1003, 1407, 1603, 1655, 1669, 1670, 1671, 1699, 1751, 1802, 1803, 1804, 1805, 1892, 1907, 2073, 2120, 2167, 2168, 2470, 2495, 2533, 2589, 2981, 3004, 3061, 3086, 3107, 3108, 3109, 3110, 3337, 3338, 3360, 3361, 3487, 3551, 3572, 3666, 3667, 3695, 3696, 3785, 3803, 3811, 3882, 3902, 3942, 3952, 3966, 3972, 3979, 4160, 4188, 4198, 4366, 4586, 4825, 5216, 5217, 5218, 5470, 5495, 5608, 5681, 5867, 5934, 5962, 5988, 6024, 6077, 6223, 6241, 6310, 6311, 6321, 6327, 6452, 6455, 6466

Cleistogenes 3814, 4174, 4328, 4980, 5048

Clematis 4613

Cleome 293, 5382, 6218

Clethra 3403

clover see *Trifolium*

Cochlearia 5329

Cocos 2533, 3144, 3145, 4637, 6072

Codiaeum 3324

Coffea 101, 102, 645, 702, 1786, 2075, 2533, 2598, 2611, 3103, 3484, 3557, 4546, 4671, 5139, 5596, 5974, 5986, 6452, 6455

Colchicum 3379

Coleus 501, 1245, 1582, 1780, 2355, 3369, 6218

Commelina 155, 648, 653, 711, 712, 1009, 1131, 1132, 1274, 1275, 1625, 1800, 1944, 1989, 2052, 2053, 2054, 2061, 2062, 2132, 2713, 2714, 3329, 3512, 3567, 3652, 3747, 4023, 4882, 5082, 5123, 5185, 5186, 5187, 5190, 5497, 5574, 5735, 5890, 5914, 5962, 5967, 5976, 5977, 5984, 5986, 5994, 6033, 6165, 6217, 6218, 6225, 6244, 6321, 6336, 6360, 6410, 6460, 6478, 6479, 6547, 6548

Comptonia 2761

Convallaria 568, 933, 2391, 2734, 4811, 5547, 6041

Convolvulus 2345, 4868, 5769, 6322

Conyza 6182

Corchorus 6218

Cordia 5588

Coreopsis 1576, 6163

Coriandrum 4738

cornel see *Cornus*

Cornus 128, 160, 763, 929, 1216, 1217, 1350, 1514, 1947, 2442, 2642, 2643, 2698, 2734, 2761, 2816, 3262 3356, 3911, 4064, 4110, 4229, 4367, 4657, 4721, 4722, 4724, 4725, 4732, 5425, 5547, 5550, 5727, 5744, 5745, 5962, 6146, 6254, 6255, 6321

Coronilla 5202, 6598

Corydalis 4768

Corylus 4499, 6312, 6313

Cosmos 1576

Cotoneaster 763, 2514, 4611, 4976

cotton see *Gossypium*

cowpea see *Vigna*

Crambe 5769

cranberry see *Vaccinium*

Crassula 1289, 2430, 4518, 4611, 5008

Crataegus 929, 1021, 1947, 2734, 4064, 4499, 4732, 4768, 4853, 5745

Crepis 3740

Crithmum 3654

Crotolaria 1590, 2941, 3083, 5572, 5804, 6163, 6218, 6321, 6322

Croton 2691, 3324, 5265, 6218

Cryptocarya 2103, 4151, 4767

Cryptomeria 3058, 5768, 5943

Craptostegia 1185

cucumber see *Cucumis*

Cucumis 140, 254, 536, 769, 805, 815, 1035, 1264, 1482, 1532, 1638, 1853, 1888, 2083, 2096, 2099, 2101, 2223, 2187, 2349, 2357, 2425, 2430, 2545, 2636, 2661, 2718, 2770, 2773, 2809, 3367, 3383, 3561, 3665, 3843, 3988, 4017, 4148, 4234, 4235, 4378, 4385, 4386, 4401, 4435, 4477, 4483, 4818, 4875, 4932, 5001, 5212, 5213, 5314, 5517, 5776, 6171, 6218, 6265, 6440, 6542

Cucurbita 627, 1121, 1236, 1625, 1797, 1887, 1910, 2092, 2128, 2141, 2880, 3695, 3696, 3843, 4148, 4448, 4875, 5111, 5509, 5695, 5776, 6225, 6386, 6424

Cuminum 6358

Curatella 3393

Curcuma 702

Cyclamen 1535

Cydonia 3737

Cymbopogon 2544, 5622

Cynodon 3, 114, 123, 1273, 1383, 1384, 2093, 2238, 3195, 3694, 4377, 4779, 5172, 5202, 5423

Cynosurus 4240

Cyperus 155, 984, 1933, 4355, 5421, 5622, 5942, 6037, 6051

Cytisus 3977, 4768, 5804, 6321

Eragrostis 2874, 2875, 4289, 5202, 5423, 5466, 5622, 5623, 6195, 6497

Erianthus 1862

Erica 1838

Erigeron 3855

Erigonum 3012

Eriophorum 845, 1851, 5278, 5395, 5711, 6027, 6108, 6464

Erodium 263, 2469, 3319, 4665

Erophyla 5077

Eryngium 1721, 5769

Erysimum 4329

Erythrina 2138, 5570

Eschscholtzia 1893, 5574

Eucalyptus 326, 471, 978, 1030, 1038, 1240, 1258, 1711, 1728, 1735, 1839, 1990, 2207, 2208, 2245, 2280, 2562, 2659, 2665, 3034, 3103, 3104, 3125, 3263, 3437, 3892, 3933, 3934, 4054, 4608, 5195, 5200, 5280, 5403, 5409, 5434, 5757, 5792, 6477, 6494, 6568, 6577

Euonymus 332, 929, 1780, 2471, 2734, 3688, 3737, 4005, 4064, 4182, 4506, 5477, 5547, 5550, 5745, 5809, 6490

Eupatorium 3728, 4056, 6039

Euphorbia 736, 1185, 1212, 1428, 1721, 1893, 2430, 2691, 3322, 3324, 3332, 4768, 4779, 4815, 4975, 5499, 5572, 5889, 6218, 6322

Eurotia 1142, 1562, 2072, 6009

Eutrema 1893

F

Fagara 2951

Fagopyrum 2094, 2859, 4197, 5672

Fagus 117, 170, 1018, 1037, 1146, 1362, 1363, 1779, 1812, 1951, 2121, 2254, 2761, 2767, 3554, 3588, 3743, 4182, 4444, 4445, 4499, 4768, 5037, 5038, 5287, 5555, 5809, 6125

Fatsia 3026

fern see *Pteridophyta*

Ferocactus 3062, 3205, 6495

Ferula 6428

fescue see

Festuca 168, 275, 383, 389, 390, 409, 603, 649, 780, 782, 897, 1005, 1066, 1312, 1383, 1457, 1539, 1625, 1757, 2253, 2596, 2844, 2870, 2953, 3005, 3195, 3214, 3379, 3380, 3433, 3550, 3719, 3822, 3856, 3868, 3943, 4240, 4279, 4328, 4347, 4405, 4456, 4468, 4471, 4473, 4474, 4504, 4577, 4655, 4768, 4779, 4811, 4867, 4873, 4881, 4937, 5077, 5180, 5181, 5202, 5381, 5389, 5423, 5554, 5611, 5628, 5629, 5659, 5732, 5734, 5871, 6165, 6243, 6307, 6351, 6352, 6353, 6428, 6443, 6531, 6539

Ficaria 6531

Ficus 815, 1185, 2471, 3103, 3384, 5139, 5777

Filipendula 3728, 3782, 4611, 6054

fir see *Abies*

Flacourtia 2851

Foeniculum 4738

Forsythia 4536, 4655, 5321

Fragaria 269, 285, 492, 530, 820, 1001, 1002, 1436, 1438, 1439, 1514, 1770, 1771, 1783, 1887, 1925, 2734, 3065, 3728, 3905, 4064, 4166, 4192, 4762, 4769, 5108, 5109, 5362, 5547

Fraxinus 129, 159, 160, 161, 321, 824, 1169, 1350, 1394, 1599, 1812, 1844, 1908, 1909, 2019, 2339, 2586, 2588, 2603, 2605, 2637, 2698, 2761, 2787, 3009, 3115, 3262, 3263, 3737, 4110, 4434, 4485, 4768, 5003, 5287, 5388, 5555, 5712, 5880, 5881

Freesia 1014

French bean see *Phaseolus*

Fumana 4779

Fungi
    Albugo 5170
    Alternaria 1880, 1982, 2000, 2275, 2395, 6030, 6036
    Aphanomyces 2135, 4189
    Armillaria 1852

Halostachys 1862

Haloxylon 1292, 1471, 2293, 4810, 5048, 6035, 6427

Hammada 223, 401, 767, 1646, 2175, 2677, 3083, 4466, 4467

Haplopappus 2347, 4131

Hedera 815, 1832, 2025, 2767, 3437, 3525, 3770, 4064, 4768, 5036, 5310, 5382, 5550, 5745, 5809, 5811, 6190, 6417

Hedysarum 5693, 5694

Heleocharis 468, 5285

Helianthus 2, 41, 43, 54, 55, 116, 229, 307, 343, 486, 566, 571, 576, 577, 600, 652, 715, 732, 810, 879, 938, 998, 1031, 1067, 1068, 1074, 1117, 1121, 1131, 1134, 1249, 1322, 1348, 1364, 1492, 1625, 1645, 1657, 1807, 1809, 1828, 1857, 1934, 2071, 2146, 2187, 2222, 2243, 2317, 2365, 2390, 2399, 2406, 2407, 2408, 2476, 2498, 2528, 2581, 2597, 2679, 2716, 2717, 2741, 2753, 2757, 2763, 2834, 2861, 2877, 2974, 3041, 3049, 3050, 3127, 3128, 3140, 3336, 3352, 3387, 3397, 3462, 3464, 3472, 3564, 3661, 3720, 3739, 3749, 3857, 3914, 3962, 3967, 4030, 4044, 4051, 4066, 4088, 4096, 4180, 4249, 4305, 4346, 4381, 4384, 4398, 4468, 4481, 4567, 4598, 4608, 4618, 4619, 4620, 4621, 4626, 4718, 4791, 4816, 4930, 4943, 5022, 5073, 5104, 5109, 5212, 5359, 5260, 5271, 5283, 5336, 5337, 5338, 5389, 5451, 5504, 5542, 5556, 5569, 5602, 5603, 5659, 5674, 5682, 5807, 5820, 5919, 5920, 6062, 6134, 6205, 6230, 6260, 6321, 6358, 6410, 6494, 6430, 6495, 6498, 6506, 6137, 6556, 6605

Helichrysum 580, 1838

Helictotrichon 4328

Heliotropium 3765, 4808

Helipterum 580

Helleborus 4768

hemp see *Cannabis*

Heracleum 1545, 6054

Heteromeles 326, 679, 1850, 3012, 5642

Heteromorpha 1832

Heteropogon 647, 6563

Heterospermum 1576

Hevea 1185. 2186, 3222, 4891, 4921, 4922, 6294, 6295

Hibiscus 991, 992, 1394, 1937, 2317, 3325, 4204, 5821

Hieracium 1457, 3005, 3006, 4768, 5871

Hilaria 1142

Hippeastrum 3642

Hippophaë 5627, 5721

Holcus 3416, 4456, 4504

holly see *Ilex*

Hordeum 31, 40, 43, 70, 71, 78, 154, 162, 212, 213, 240, 247, 261, 264, 282, 283, 300, 316, 386, 427, 453, 507, 592, 593, 594, 690, 691, 711, 712, 723, 804, 846, 851, 968, 1000, 1022, 1023, 1051, 1074, 1096, 1097, 1111, 1113, 1114, 1117, 1134, 1226, 1227, 1232, 1261, 1285, 1306, 1324, 1365, 1393, 1398, 1415, 1434, 1481, 1487, 1488, 1511, 1556, 1589, 1619, 1683, 1707, 1724, 1773, 1790, 1809, 1837, 1928, 1931, 1934, 1955, 2035, 2039, 2040, 2041, 2089, 2092, 2097, 2100, 2107, 2128, 2204, 2225, 2311, 2317, 2321, 2367, 2388, 2404, 2415, 2434, 2448, 2468, 2475, 2486, 2523, 2524, 2558, 2596, 2645, 2700, 2722, 2740, 2752, 2759, 2767, 2770, 2824, 2885, 2942, 2949, 2956, 2957, 2963, 2972, 3014, 3085, 3087, 3101, 3121, 3123, 3143, 3162, 3183, 3185, 3196, 3264, 3270, 3295, 3300, 3301, 3302, 3303, 3316, 3335, 3336, 3367, 3375, 3412, 3420, 3425, 3447, 3458, 3463, 3469, 3510, 3515, 3545, 3559, 3592, 3609, 3674, 3747, 3749, 3777, 3801, 3805, 3926, 3950, 3985, 4017, 4045, 4069, 4100, 4101, 4178, 4183, 4210, 4217, 4218, 4219, 4228, 4273, 4281, 4290, 4329, 4332, 4381, 4406, 4426, 4430, 4504, 4537, 4623, 4646, 4647, 4679, 4760, 4781, 4782, 4783, 4789, 4804, 4805, 4806, 4839, 4840, 4841, 4842, 4843, 4906, 4910, 4923, 4924, 4952, 5023, 5024, 5028, 5031, 5088, 5101, 5103, 5110, 5116, 5117, 5186, 5206, 5207, 5208, 5221, 5222, 5223, 5252, 5283, 5289, 5290, 5465, 5469,

Hordeum (continued) 5500, 5504, 5510, 5511, 5594, 5624, 5645, 5659, 5686, 5698, 5699, 5772, 5773, 5861, 5898, 5903, 5918, 5919, 5923, 5928, 5939, 5960, 5986, 6023, 6048, 6067, 6135, 6192, 6358, 6367, 6372, 6395, 6396, 6397, 6408, 6447, 6452, 6453, 6455, 6491, 6492, 6494, 6514, 6523, 6574, 6586

hornbeam see *Carpinus*

Hoya 4678

Humulus 1968, 3870, 4136, 4715, 4732, 4738, 6253

Hyparrhenia 1064

Hypochoeris 1938, 2207, 4038, 4456

Hyptis 2255, 2469, 3395, 3479, 3480, 6218, 6386

Hyssopus 4738

I

Iberis 4811

Ilex 195, 815, 1635, 1953, 3356, 3748, 4861, 5153, 5712, 5809, 6254, 6255, 6551

Impatiens 1514, 1625, 2317

Imperata 1862, 4180, 6051

Inula 3654, 5769, 6523

Ipomoea 323, 669, 747, 904, 1276, 1625, 1966, 1976, 2141, 3399, 4751, 5041, 6077, 6322

Iris 1625, 5049

J

Jatropha 3324, 6218

Juglans 117, 160, 339, 824, 1198, 1514, 1711, 1812, 1855, 2578, 2698, 3430, 3780, 3807, 3831, 3898, 4192, 4681, 4819, 4859, 5115, 5555, 5597, 5727, 6452, 6455

Juncus 468, 984, 2124, 2125, 3668, 3704, 4157, 5711, 5942, 6395

Juniperus 1169, 1350, 1389, 1741, 1852, 1940, 2317, 2603, 3832, 5153, 5388, 5393, 5428, 5727, 5943, 6551

K

Kalanchoë 259, 509, 591, 970, 1078, 1289, 1591, 1625, 1946, 2327, 2640, 2641, 3080, 3279, 3660, 4527, 5572, 5781, 5854, 6218

Kalmia 5297

Knema 2226

Kobresia 199, 4328, 5395

Kochia 123, 1471, 2056, 4665, 4810, 5127, 5769, 5791, 6428

Koeleria 275, 2072, 4328, 4830, 6246, 6268, 6428

Kosteletzkya 6395, 6396

Krameria 46

L

Lablab 2590

Lactuca 7, 140, 214, 724, 793, 995, 1202, 1317, 1829, 1888, 1893, 2007, 2110, 2225, 2286, 2317, 2332, 2487, 2546, 2673, 2732, 2819, 2923, 3179, 3443, 3469, 3561, 3695, 3696, 3697, 3855, 3883, 3891, 3990, 4645, 4933, 5013, 5039, 5357, 5407, 5408, 5517, 5716, 6265, 6292, 6556

Lagenaria 361, 4971

Lagerstroemia 4629

Laguncularia 556

Lamium 2734, 4064, 5547

Landolphia 1185

larch see *Larix*

Larix 547, 909, 1037, 1175, 1250, 1852, 2317, 2322, 2762, 3554, 3787, 4182, 4239, 4328, 5049, 5768, 5943, 6477, 6494

Larrea 46, 868, 1134, 1169, 2603, 2767, 3160, 4606, 4611, 5051, 5318, 5388, 5453, 6009, 6049, 6291, 6297

Lasthenia 6395

Lathyrus 2083, 3379, 4877, 4878, 5327, 5572

Laurus 1999, 2375, 5572

Lavandula   1838, 4738

Lavatera 2509

Ledum 705

Lemna 1065, 1827, 2399, 2814, 5312, 5737, 5865

Lens 2138, 3118, 3463, 4452, 5651, 6329

Leontopodium 275

Lepidium 3704, 4242, 4243, 6556

Lepturus 6523

Lespedeza 1705, 5202, 5804, 6321

lettuce see *Lactuca*

Levisticum 4738

Libocedrus 1852

Lichenes
    Alectoria 412, 482, 483, 484,
    1738, 2279, 3045, 4475, 4613,
    5915
    Buellia 5102
    Candelariella 5102
    Cetraria 482, 484, 1738, 2279,
    3006, 4613, 5042
    Cladina 412, 413, 1738, 5915,
    5925
    Cladonia 3006, 3045, 3876,
    3894, 4145, 4470, 4613, 5042,
    5102, 5384, 5926
    Dermatocarpon 5384
    Evernia 3876, 5102, 5384
    Hypogymnia 1358, 1359, 3876,
    5102, 5384
    Lecanora 5102
    Lepraria 5102
    Leptogium 5384
    Parmelia 482, 1025, 2512,
    3876, 5102, 5384, 5832
    Peltigera 482, 1357, 2990,
    3876, 4074, 4145, 4393, 4394,
    5384, 5832
    Pertusaria 5102
    Phlyctis 5102
    Physcia 1025, 2512
    Physconia 5102
    Platismatia 5102, 5384
    Ramalina 3042, 5324, 5384,
    5806, 6342
    Rhizocarpon 3276
    Stereocaulon 4295, 4296, 4395,
    5449
    Sticta 6277
    Teloschistes 6342
    Thamnolia 3006, 3045

Lichenes
    Umbilicaria 3045, 3276, 5384,
    5915, 5916
    Usnea 3876, 5102, 5384
    Xanthoria 1357, 1887, 5102

Ligustrum 2734, 4064, 5153, 5547,
5550, 5745, 5871, 5902, 5990

Lilium 4616, 4768, 6090, 6091

Limnanthes 3881

Limonium 5278, 5770, 6428

Lynosyris 6428

Linum 92, 93, 1322, 1853, 2225, 3085,
3128, 3703, 4196, 5259, 5260, 6014,
6452, 6455

Liquidambar 146, 763, 2602, 3356,
5555

Liriodendron 1755, 1812, 2094, 2119,
2602, 3356, 3566, 4530, 4569, 5061,
5555, 6255

loblolly pine see *Pinus*

Loiseleuria 1701, 1702, 2631, 2632,
2841, 2842, 3005, 3006, 5711

Lolium 124, 142, 168, 262, 312, 383,
409, 666, 694, 780, 781, 782, 955,
956, 1063, 1066, 1074, 1243, 1253,
1298, 1383, 1661, 1747, 1885, 1895,
1932, 2009, 2074, 2128, 2203, 2253,
2301, 2315, 2447, 2596, 2657, 2789,
2870, 2876, 2894, 2895, 2994, 3038,
3214, 3258, 3416, 3433, 3550, 3657,
3703, 3783, 3835, 3925, 3956, 4079,
4224, 4225, 4227, 4240, 4343, 4347,
4354, 4456, 4474, 4732, 4872, 4873,
4937, 4939, 4993, 5077, 5079, 5103,
5150, 5188, 5197, 5202, 5321, 5375,
5389, 5423, 5466, 5499, 5599, 5628,
5629, 5714, 5725, 5726, 5861, 5953,
6156, 6168, 6225, 6242, 6269, 6443,
6483, 6513, 6531, 6575, 6579

Lomatium 2072

Lonicera 1021, 1187, 2024, 4110,
4499, 6428

Loranthus 5551

Lotononis 1663

Lotus 1308, 2138, 2790, 2791, 3214,
3416, 4022, 4117, 4118, 4937, 5439,
6598

Luffa 4971

Lunaria 5382

lupine see *Lupinus*

Lupinus 68, 840, 2138, 2266, 2590,
2591, 3040, 3307, 3943, 4256, 4257,
4258, 4259, 4271, 4464, 4728, 5270,
5804, 5919, 6321, 6351, 6352, 6353,
6556, 6569, 6570, 6571, 6572, 6573

Luzula 4456, 5871

Lycium 46, 1862, 3733

Lycopersicon 1, 41, 89, 124, 179,
214, 220, 260, 366, 442, 536, 559,
560, 583, 614, 681, 747, 748, 790,
793, 806, 846, 903, 982, 988, 1036,
1059, 1074, 1075, 1097, 1117, 1134,
1160, 1245, 1268, 1322, 1339, 1424,
1459, 1462, 1514, 1521, 1604, 1605,
1627, 1634, 1638, 1800, 1814, 1856,
1866, 1874, 1887, 1910, 1934, 1975,
1982, 2066, 2083, 2094, 2096, 2128,
2141, 2176, 2187, 2305, 2317, 2365,
2463, 2471, 2517, 2535, 2545, 2627,
2628, 2669, 2724, 2725, 2960, 2991,
3123, 3162, 3178, 3268, 3282, 3284,
3290, 3312, 3367, 3374, 3399, 3470,
3532, 3607, 3645, 3722, 3747, 3760,
3784, 3877, 3897, 3945, 3991, 4000,
4069, 4106, 4181, 4246, 4264, 4308,
4331, 4332, 4357, 4384, 4424, 4448,
4560, 4584, 4594, 4615, 4632, 4633,
4758, 4865, 4870, 4871, 4910, 4920,
4964, 4969, 4970, 5064, 5065, 5066,
5071, 5072, 5074, 5118, 5146, 5212,
5234, 5268, 5269, 5281, 5286, 5382,
5389, 5411, 5504, 5517, 5572, 5654,
5678, 5729, 5735, 5764, 5810, 5836,
5866, 5894, 5953, 5962, 5971, 5999,
6003, 6012, 6015, 6024, 6073, 6185,
6244, 6321, 6396, 6434, 6435, 6440,
6447, 6452, 6455, 6494, 6499, 6512,
6544

Lysimachia 1005, 1747, 3728, 4456,
4732, 6156

M

Macropiper 3211

Magnolia 117, 1812, 4367, 4506

Maianthemum 3728, 3761

maize see *Zea*

Malus 28, 36, 57, 81, 217, 270, 294,
344, 357, 358, 359, 444, 459, 462,

Malus (continued) 463, 466, 474,
608, 615, 642, 677, 686, 820, 832,
864, 1184, 1214, 1231, 1257, 1350,
1368, 1477, 1533, 1569, 1673, 1697,
1732, 1751, 1899, 1934, 1977, 1997,
2017, 2024, 2029, 2030, 2031, 2136,
2177, 2316, 2421, 2422, 2540, 2571,
2904, 3041, 3105, 3253, 3294, 3296,
3306, 3771, 3908, 3911, 3958, 3974,
3975, 4082, 4093, 4122, 4169, 4170,
4171, 4216, 4434, 4450, 4457, 4458,
4554, 4611, 4955, 5067, 5081, 5084,
5160, 5165, 5166, 5168, 5233, 5288,
5301, 5302, 5467, 5468, 5589, 5677,
5678, 5763, 5786, 5885, 5901, 5905,
5934, 5961, 5963, 5964, 6048, 6086,
6146, 6275, 6385, 6400, 6401, 6482

Mamillaria 5087

mangrove see *Rhizophora*

Magnifera 2261, 6362

Manihot 2515, 3092, 3093, 4300,
4301, 6083, 6218

maple see *Acer*

Maranta 6410

Marrubium 5769

Matricaria 3044, 4738

Medicago 103, 346, 372, 374, 375,
392, 532, 647, 921, 937, 951, 964,
1049, 1078, 1082, 1102, 1103, 1233,
1245, 1308, 1312, 1373, 1392, 1451,
1489, 1490, 1519, 1539, 1586, 1593,
1625, 1658, 1695, 1718, 1755, 1825,
1869, 1895, 1896, 1924, 1963, 1995,
2083, 2094, 2097, 2247, 2301, 2317,
2524, 2525, 2624, 2692, 2781, 2789,
2882, 2883, 2906, 2924, 2938, 3014,
3015, 3021, 3136, 3137, 3138, 3183,
3308, 3309, 3310, 3357, 3416, 3467,
3469, 3472, 3545, 3638, 3651, 3718,
3769, 3781, 3794, 3817, 3821, 3948,
3950, 3998, 4017, 4022, 4050, 4079,
4086, 4090, 4101, 4117, 4118, 4167,
4245, 4269, 4277, 4388, 4421, 4501,
4507, 4545, 4612, 4713, 4718, 4740,
4779, 4795, 4796, 4797, 4798, 4810,
4926, 4937, 4988, 5037, 5038, 5091,
5272, 5299, 5321, 5355, 5385, 5439,
5486, 5494, 5651, 5678, 5685, 5769,
5804, 5831, 5921, 5938, 5955, 6011,
6169, 6179, 6211, 6212, 6213, 6214,
6238, 6320, 6321, 6334, 6350, 6378,
6443, 6452, 6453, 6454, 6455, 6458,
6494, 6598

Melia 4367

Persica 269, 462, 466, 608, 698, 726, 1161, 1436, 1437, 1555, 1913, 2559, 2644, 3147, 3343, 3677, 3906, 4156, 4192, 4450, 4603, 4640, 4792, 4847, 5139, 5225, 5293, 5678, 5896, 6385, 6588

Petroselinum 4738

Petunia 1284, 2331, 2825, 2826, 3315, 3399, 4163, 5112, 5552, 5553, 5644

Phalaris 123, 383, 1312, 1354, 1383, 1387, 2152, 2253, 2315, 2596, 2835, 3379, 3380, 3821, 4159, 4220, 4347, 4881, 4937, 5085, 5296, 5466, 5628, 5629, 5653, 6243

Pharbitis 3026

Phaseolus 11, 30, 32, 50, 121, 151, 205, 214, 250, 251, 322, 367, 378, 379, 417, 418, 461, 464, 465, 467, 501, 537, 579, 616, 617, 634, 664, 675, 678, 694, 701, 711, 712, 743, 756, 774, 787, 790, 796, 810, 827, 828, 835, 853, 903, 949, 980, 1042, 1055, 1056, 1097, 1099, 1134, 1145, 1154, 1160, 1209, 1245, 1269, 1340, 1397, 1414, 1442, 1492, 1500, 1507, 1514, 1550, 1609, 1617, 1621, 1625, 1628, 1659, 1726, 1749, 1755, 1792, 1793, 1795, 1796, 1809, 1818, 1832, 1847, 1860, 1869, 1875, 1887, 1888, 1893, 1910, 1934, 1948, 1984, 1999, 2011, 2021, 2027, 2028, 2083, 2094, 2115, 2128, 2155, 2170, 2185, 2210, 2218, 2225, 2227, 2256, 2264, 2266, 2295, 2317, 2370, 2374, 2428, 2438, 2446, 2457, 2458, 2459, 2460, 2590, 2699, 2718, 2726, 2753, 2785, 2928, 2938, 2952, 3037, 3055, 3072, 3128, 3149, 3154, 3156, 3157, 3169, 3183, 3185, 3186, 3216, 3218, 3219, 3239, 3249, 3250, 3259, 3271, 3272, 3291, 3292, 3293, 3357, 3377, 3488, 3489, 3490, 3514, 3568, 3570, 3625, 3627, 3633, 3663, 3695, 3696, 3737, 3890, 3945, 3962, 3982, 3995, 4063, 4084, 4115, 4123, 4124, 4148, 4168, 4180, 4274, 4276, 4303, 4333, 4335, 4342, 4376, 4396, 4429, 4439, 4450, 4451, 4472, 4544, 4560, 4600, 4637, 4642, 4673, 4680, 4723, 4736, 4754, 4772, 4773, 4774, 4778, 4802, 4876, 4901, 4914, 4925, 4930, 4931, 4938, 4996, 5014, 5083, 5125, 5139, 5186, 5212, 5279, 5313, 5315, 5321, 5345, 5379, 5389, 5446, 5463, 5472, 5474, 5476, 5488, 5490, 5504, 5509, 5517, 5595, 5610, 5633, 5678, 5680, 5686, 5694, 5702, 5706, 5730, 5735, 5804, 5807, 5884, 5914, 5952, 6005, 6041, 6076,

Phaseolus (continued) 6112, 6116, 6139, 6248, 6253, 6274, 6306, 6321, 6392, 6429, 6452, 6455, 6467, 6494, 6504, 6508, 6513, 6521, 6526, 6530, 6549, 6554, 6556, 6582

Phelline 1041

Phellodendron 2509

Philadelphus 6490

Phillyrea 2375

Philodendron 3946

Philoxeros 6182

Phleum 782, 1066, 1539, 1757, 3214, 3416, 4240, 4546, 4577, 5254, 5499, 5628, 5629

Phoenix 6452

Phragmites 455, 1081, 1862, 3552, 3704, 4159, 4432, 4433, 4479, 4732, 4880, 5285, 5942, 6156, 6428

Phyllanthus 2691, 4161

Physalis 1974, 6218

Physocarpus 1235, 3262

Picea 137, 163, 279, 290, 406, 511, 514, 601, 651, 739, 815, 877, 911, 934, 936, 1037, 1045, 1115, 1175, 1250, 1461, 1508, 1510, 1525, 1654, 1660, 1665, 1692, 1820, 1821, 1852, 2013, 2044, 2078, 2128, 2322, 2392, 2417, 2518, 2541, 2582, 2585, 2587, 2652, 2678, 2854, 2932, 2980, 3041, 3256, 3289, 3355, 3378, 3381, 3547, 3554, 3658, 3744, 3773, 3774, 3840, 3879, 3907, 3954, 3955, 3969, 4026, 4194, 4239, 4250, 4251, 4444, 4488, 4528, 4548, 4652, 4750, 4860, 5037, 5038, 5106, 5152, 5158, 5311, 5373, 5428, 5491, 5605, 5680, 5735, 5768, 5775, 5809, 5811, 5899, 5943, 6031, 6063, 6065, 6321, 6477, 6494, 6514, 6531

Pimpinella 4738, 6431

pine see *Pinus*

Pinus 5, 100, 114, 159, 225, 257, 284, 335, 345, 377, 395, 517, 599, 619, 639, 651, 729, 739, 815, 841, 842, 855, 866, 961, 1034, 1037, 1038, 1045, 1048, 1072, 1080, 1134, 1138, 1162, 1167, 1228, 1235, 1258, 1289, 1309, 1338, 1350, 1352, 1370, 1403,

Pinus (continued) 1430, 1443,
1514, 1525, 1529, 1530, 1575, 1602,
1616, 1635, 1711, 1781, 1787, 1820,
1939, 2048, 2083, 2094, 2109, 2113,
2128, 2149, 2150, 2151, 2184, 2257,
2296, 2322, 2323, 2377, 2380, 2440,
2555, 2557, 2569, 2601, 2602, 2603,
2605, 2609, 2652, 2666, 2675, 2678,
2793, 2829, 2852, 2854, 2900, 2979,
2980, 2981, 2999, 3071, 3125, 3139,
3189, 3254, 3256, 3257, 3260, 3261,
3343, 3354, 3364, 3370, 3373, 3378,
3391, 3481, 3548, 3554, 3558, 3599,
3603, 3635, 3647, 3762, 3763, 3787,
3790, 3791, 3887, 3907, 3955, 4075,
4137, 4141, 4164, 4182, 4239, 4251,
4328, 4380, 4419, 4434, 4443, 4445,
4446, 4488, 4579, 4630, 4652, 4741,
4750, 4821, 4895, 4955, 5007, 5026,
5037, 5038, 5046, 5049, 5159, 5215,
5270, 5278, 5373, 5388, 5393, 5401,
5405, 5448, 5509, 5559, 5563, 5568,
5597, 5689, 5703, 5721, 5747, 5762,
5768, 5883, 5899, 5943, 5944, 6092,
6138, 6180, 6239, 6279, 6388, 6420,
6477, 6494, 6514, 6517, 6532, 6597

Piper 214, 442, 793, 2223, 2599,
3607, 3695, 3829, 2862, 3877, 4637

Pirus 217, 286, 462, 465, 466,
1726, 2167, 2745, 2746, 2850, 2887,
3025, 3028, 3265, 4073, 4192, 4299,
4434, 4450, 4611, 5139, 5168, 5210,
5211, 5819, 6146, 6290, 6583

Pistacia 2375, 4329, 4810, 6551

Pisum 32, 104, 109, 110, 155, 216,
267, 296, 520, 558, 609, 743, 757,
793, 799, 810, 839, 853, 943, 971,
972, 1040, 1136, 1172, 1181, 1186,
1366, 1487, 1598, 1625, 1818, 1887,
1893, 1971, 1999, 2000, 2011, 2092,
2094, 2146, 2170, 2220, 2290, 2317,
2345, 2396, 2406, 2476, 2521, 2522,
2664, 2716, 2748, 2749, 2750, 2782,
2789, 2872, 2947, 2995, 3014, 3102,
3128, 3150, 3362, 3371, 3522, 3537,
3610, 3684, 3694, 3737, 3741, 3747,
3768, 3786, 3940, 3945, 4539, 4556,
4560, 4653, 4684, 4704, 4777, 4852,
4876, 4930, 4955, 5055, 5205, 5232,
5278, 5303, 5382, 5402, 5503, 5658,
5702, 5738, 5738, 5761, 5769, 5804,
5808, 5864, 5884, 5919, 5967, 5994,
6193, 6202, 6321, 6410, 6452, 6455,
6520, 6543, 6556, 6598

Pittosporum 2207, 2208, 3437, 5902

Plantago 148, 501, 1105, 1289, 1625,
1882, 1938, 2131, 3175, 3654, 3704,
4738, 5255, 5329, 5332, 5583, 5586,
5650, 5770, 5771, 6442

Platanus 824, 2375, 2622, 3599,
4492, 4540, 5061, 5368, 6598

Platycodon 4811

Plectranthus 3206

Poa 223, 285, 780, 940, 1005, 1066,
1312, 2317, 2915, 3214, 4011, 4038,
4240, 4328, 4329, 4343, 4456, 4474,
4655, 4732, 4768, 5049, 5077, 5202,
5423, 5439, 5628, 5629, 5653, 5871,
6243, 6351, 6352, 6353

Podocarpus 1917, 1918, 3211

Polygonatum 2734, 3728, 4064, 4504,
4768, 5547, 6531

Polygonum 1893, 1966, 2008, 2272,
2432, 3855, 4752, 4753, 5572, 5712

Polypogon 6523

pomegranate see *Punica*

Populus 285, 339, 485, 670, 671, 720,
729, 824, 1037, 1134, 1187, 1281,
1282, 1307, 1330, 1362, 1403, 1445,
1596, 1703, 1739, 1779, 1862, 2023,
2024, 2083, 2339, 2509, 2595, 2686,
2693, 2761, 2787, 3046, 3263, 3437,
3587, 3599, 3625, 3678, 3903, 3904,
4109, 4110, 4182, 4252, 4310, 4434,
4476, 4488, 4732, 4761, 4808, 4901,
4944, 4955, 5006, 5106, 5237, 5239,
5386, 5412, 5414, 5462, 5463, 5506,
5507, 5600, 5696, 5721, 5768, 6130,
6131, 6371, 6597, 6598

Portulaca 1630, 2962, 2987, 3332,
3855, 4359, 4420, 5499, 6231, 6513

Posidonia 4447

Potamogeton 2167, 3716, 4096

potato see *Solanum*

Potentilla 275, 468, 845, 1105,
5499, 5871, 6027

Primula 2633, 3005, 3006, 4471

Prosopis 193, 271, 773, 1110, 1134,
1337, 2931, 3414, 3576, 5462. 5463.
5743, 6322

Protium 2961

Prunella 1938

Prunus 188, 217, 218, 294, 462, 466,
651, 685, 698, 794, 862, 1126, 1178,
1211, 1362, 1394, 1751, 1812, 2160,

Spirodela 5196

Sporobolus 984, 1064, 2739, 3816, 4349, 5294, 5623, 6351, 6352, 6353, 6497

spruce see *Picea*

Stachytarpheta 1986, 1987, 3255, 6163, 6164, 6165, 6166

Stangeria 2377

Statice 1136

Stellaria 929, 2317, 4456, 4504, 4768, 50.77, 5871

Sterculia 4629

Stipa 275, 741, 822, 2195, 2293, 2294, 2876, 3067, 4174, 4328, 4665, 4980, 5048, 5049, 5423, 5499, 5622, 6056, 6268, 6351, 6352, 6353, 6428

Stizolobium 6165

strawberry see *Fragaria*

Strophanthus 5098

Stylidium 4098, 4099

Stylosanthes 1049, 1190, 2116, 2346, 2525, 3066, 3129, 4058, 4104, 4568, 5686, 5803, 5804, 6321, 6495

Suaeda 123, 846, 1471, 3664, 3704, 3833, 4107, 4108, 4209, 5208, 5330, 5331, 5359, 5360, 5369, 5499, 5722, 5962, 6396, 6416

sugar beet see *Beta*

sugar cane see *Saccharum*

sweet potato see *Ipomoea*

Symphoricarpus 5721

Synedrella 1576

Syringa 197, 1350, 4499, 5061

Syzigium 5889, 5902

T

Tagetes 4163, 6039

Talbotia 6497

Tamarix 1281, 1282, 1530, 1637, 1843, 1844, 1862, 2693, 5462, 5463, 5962, 6321

Tanacetum 4639, 6428

Taraxacum 1005, 1185, 1798, 2207, 2317, 2986, 3026, 3437, 4456, 5077, 5442, 5871

Taverniera 1705, 6321

Taxodium 5943, 6038

Taxus 2377, 4182, 4655

Tectona 1592, 4304, 4629

Terminalia 4629, 5889, 5951

Tetradymia 1142

Teucrium 4617, 5769, 5813

Thalassia 2173

Thalictrum 1389, 6428

Thea 233, 997, 1203, 1372, 1934, 1957, 2623, 2733, 3055, 3893, 4059, 4255

Thelesperma 1576

Theobroma 25, 26, 489, 1610, 2533, 3213, 4637, 6452, 6455

Thuja 587, 1382, 1852, 3368, 4182, 4291, 4737, 5428, 5607, 5762, 5943

Thujopsis 2377, 5943

Thymus 4738, 4779, 5049, 5877

Thyridolepis 133, 134

Tidestromia 65, 75, 4818, 5318, 5499, 6297

Tilia 1403, 1779, 1812, 1999, 2509, 2686, 3009, 4182, 4434, 4499, 4768, 5287, 5393, 5555, 6494

Tillandsia 1301, 2386, 4571, 5137, 5909

tobacco see *Nicotiana*

tomato see *Lycopersicon*

Tradescantia 196, 293, 524, 539, 540, 786, 1189, 1329, 1830, 1832, 2206, 2471, 2716, 2731, 3079, 3437, 3438,

Triticum (continued) 5919, 5920,
5923, 5956, 5960, 5985, 5996, 6012,
6018, 6019, 6042, 6069, 6095, 6101,
6105, 6106, 6110, 6127, 6135, 6150,
6151, 6152, 6168, 6170, 6189, 6192,
6201, 6206, 6208, 6209, 6210, 6256,
6262, 6269, 6276, 6278, 6283, 6289,
6314, 6323, 6324, 6328, 6339, 6358,
6363, 6364, 6368, 6372, 6379, 6381,
6382, 6391, 6397, 6408, 6446, 6447,
6452, 6453, 6455, 6458, 6468, 6469,
6470, 6481, 6484, 6489, 6494, 6553,
6554, 6557, 6574, 6580, 6584, 6592

Trollius 1005, 3728

Tropaeolum 2509

Tsuga 201, 877, 913, 1026, 1812,
1852, 2128, 2776, 2799, 3368, 3849,
4182, 5106, 5373, 5833, 5943

Tulipa 223, 2061, 2062, 2373,
2714, 3329, 4023, 5574, 5967, 6033,
6321, 6360, 6428

Tussilago 5244

Typha 455, 1081, 2420, 2812, 3016,
3017, 3395, 3714, 4432, 4433, 4955,
5278, 5942

U

Ulex 2138, 2356

Ulmus 160, 920, 1043, 1834, 1844,
1908, 1947, 2023, 2024, 2698, 2761,
3262, 3263, 3353, 3747, 4499, 4650,
4683, 5367, 5386, 5555, 5880

Urtica 3728

Utricularia 867

V

Vaccinium 18, 73, 705, 1005, 1050,
1327, 1403, 2279, 2631, 4470, 4499,
5615, 5711, 5869, 5871, 5929, 6108,
6441

Valeriana 1433, 5049

Valerianella 2132, 4515, 4516

Vallisneria 464, 1131, 1726, 4449,
4450

Vanda 301

Vanilla 3191

Veratrum 3728, 5869, 5871

Verbascum 1721, 1964, 2439, 5769

Verbena 5994

Veronica 275, 1625, 1721, 1882, 3175,
4732, 4768, 5077, 5959

Viburnum 1021, 1187, 2761, 4499,
4536

Vicia 131, 189, 196, 351, 403, 404,
465, 501, 628, 711, 712, 740, 760,
800, 1005, 1009, 1131, 1192, 1322,
1349, 1373, 1382, 1485, 1582, 1625,
1650, 1773, 1800, 1809, 1832, 1842,
1883, 1887, 1893, 1910, 1949, 2061,
2094, 2096, 2111, 2132, 2138, 2139,
2146. 2147, 2163, 2208, 2244, 2266,
2317, 2342, 2405, 2526, 2624, 2716,
2733, 2770, 2773, 2814, 2969, 2970,
3055, 3076, 3077, 3123, 3139, 3162,
3371, 3437, 3438, 3453, 3494, 3495,
3597, 3622, 3631, 3669, 3724, 3880,
3976, 4023, 4059, 4130, 4206, 4210,
4261, 4370, 4371, 4372, 4373, 4381,
4389, 4412, 4448, 4517, 4525, 4546,
4608, 4694, 4695, 4792, 4823, 4874,
4876, 4877, 4878, 4907, 4908, 4940,
4945, 4949, 4955, 4991, 5015, 5110,
5122, 5123, 5339, 5340, 5343, 5344,
5441, 5497, 5652, 5680, 5735, 5769,
5804, 5919, 5962, 5966, 5967, 5976,
5977. 5978. 5994, 6033, 6041, 6070,
6123, 6124, 6165, 6194, 6198, 6199,
6225, 6258, 6272, 6321, 6337, 6410,
6452, 6455, 6494, 6556, 6598

Vigna 761, 762, 1047, 1097, 1170,
1484, 1730, 1854, 2590, 2783, 2805,
3410, 3411, 3574, 4460, 4591, 4909,
4968, 5062, 5212, 5250, 5454, 5686,
5804, 5968, 6157, 6218, 6262, 6318,
6321, 6326, 6603

Vinca 1893, 6218, 6285

Viola 1060, 1625, 1893, 2734, 4064,
5547, 6000

Viscum 4955, 6290

Vitis 27, 190, 285, 428, 450, 451,
462, 503, 516, 544, 550, 608, 696,
767, 904, 1011, 1074, 1117, 1134,
1144, 1225, 1319, 1320, 1321, 1351,
1388, 1502, 1547, 1548, 1549, 1714,
1715, 1861, 2088, 2314, 2366, 2436,
2505, 2579, 2719, 2773, 2916, 3088,
3089, 3193, 3194, 3247, 3265, 3470,
3528, 3760, 3795, 4033, 4041, 4192,
4322, 4500, 4592, 4631, 4745, 4863
4864, 4912, 5005, 5063, 5404, 5518,

SUBJECT INDEX
(Cumulative index to volumes 1 - 5)

This index contains a selection of primary items chosen according to their interest
for water relation researchers and to their relative importance and occurrence.

A

Age of plant, effect on water transport in cells 627, 1100, 1752, 2022,

Age of plant, effect on water transport in plant 651, 742, 1017, 1910, 2174, 4427,
    4428, 4985, 5595

Age of plant, effect on wilting 33, 68, 148, 162, 167, 317, 778, 1017, 1074, 1205,
    1447, 1622, 1663, 1716, 1735, 1970, 2153, 2214, 2236, 2372, 2563, 3015, 3041,
    3100, 3242, 3257, 3388, 3418, 3484, 3592, 3764, 3957, 4006, 4030, 4104, 4219,
    4260, 4270, 4370, 4727, 4816, 4836, 4857, 4893, 5033, 5225, 5933, 6006

Altitude and pressure, effect on conductance for water vapour and CO_2 transfer 6386

Altitute and pressure, effect on stomata and epidermis 6604

Altitude and pressure, effect on transpiration 387, 6386, 6477

Altitude and pressure, effect on water status in plant 2128, 3209

Altitude and pressure, effect on wilting 176

Amino acids see Proteins, amino acids, nucleic acids, ...

Amphistomatous leaves 8, 124, 260, 590, 2783, 2784, 2785, 2984, 4051, 4963

Anatomical structure, effect on conductance for water vapour and CO_2 transfer 5314

Anatomical structure, effect on transpiration 125, 3251, 3332

Anatomical structure, effect on water absorption by plant 6193, 6318

Anatomical structure, effect on water transport in plant 5882

Anatomical structure of epidermis 4003, 4744, 5314, 5365, 5972, 6037, 6225

Antibiotics, effect on transpiration 1727

Antibiotics, effect on water absorption by plant 1727, 3499

Antitranspirants (see also Growth substances, hormones, inhibitors etc., ...) 2,
    148, 156, 159, 161, 246, 247, 248, 384, 447, 448, 562, 576, 777, 811, 878, 914,
    916, 920, 973, 1002, 1003, 1074, 1097, 1257, 1269, 1281, 1282, 1348, 1350, 1393,
    1434, 1462, 1507, 1677, 1711, 1798, 1921, 1953, 1978, 1979, 2136, 2241, 2244,
    2271, 2307, 2325, 2374, 2486, 2495, 2534, 2550, 2668, 2686, 2693, 2736, 2766,
    2773, 2816, 2817, 3011, 3064, 3066, 3123, 3182, 3213, 3307, 3317, 3368, 3425,
    3633, 3646, 3666, 3834, 3896, 3907, 3917, 3919, 3965, 3978, 3980, 4066, 4093,
    4472, 4546, 4588, 4643, 4930, 4931, 4997, 5160, 5204, 5252, 5458, 5462, 5463,
    5672, 5896, 5914, 5045, 5961, 5963, 5964, 5984, 5986, 5988, 6001, 6013, 6024,
    6025, 6076, 6114, 6164, 6165, 6321, 6322, 6340, 6413, 6529

Availability of soil water 6, 34, 38, 49, 51, 100, 102, 103, 111, 114, 115, 120,
    146, 158, 159, 160, 163, 169, 173, 180, 181, 185, 188, 193, 208, 211, 213, 227,
    240, 256, 274, 281, 290, 312, 327, 350, 383, 388, 410, 436, 437, 453, 511, 518,
    521, 557, 561, 580, 585, 598, 606, 619, 633, 638, 649, 650, 676, 679, 692, 693,
    721, 725, 735, 737, 738, 776, 788, 803, 808, 840, 868, 883, 912, 941, 958,
    966, 982, 986, 994, 998, 1006, 1012, 1013, 1026, 1033, 1034, 1039, 1044, 1060,
    1064, 1069, 1077, 1079, 1096, 1109, 1137, 1140, 1143, 1166, 1175, 1176, 1194,
    1204, 1206, 1207, 1217, 1278, 1292, 1305, 1312, 1327, 1334, 1337, 1341, 1343,
    1352, 1359, 1360, 1365, 1381, 1382, 1389, 1393, 1395, 1424, 1440, 1442, 1446,
    1458, 1464, 1465, 1470, 1486, 1491, 1519, 1520, 1533, 1547, 1563, 1571, 1581,
    1584, 1585, 1588, 1601, 1606, 1608, 1610, 1612, 1615, 1616, 1627, 1642, 1645,
    1650, 1669, 1674, 1681, 1682, 1686, 1691, 1694, 1695, 1723, 1729, 1736, 1740,
    1743, 1744, 1748, 1749, 1759, 1770, 1781, 1789, 1804, 1810, 1812, 1814, 1816,
    1846, 1852, 1859, 1870, 1872, 1884, 1895, 1896, 1897, 1914, 1957, 1966, 1978,
    2001, 2002, 2004, 2005, 2009, 2010, 2023, 2024, 2038, 2042, 2050, 2051, 2074,

Availability of soil water (continued) 2079, 2080, 2083, 2084, 2091, 2093, 2112,
    2134, 2151, 2165, 2180, 2186, 2195, 2198, 2203, 2212, 2220, 2232, 2236, 2237,
    2246, 2247, 2253, 2260, 2266, 2274, 2276, 2280, 2285, 2288, 2292, 2296, 2302,
    2306, 2307, 2314, 2326, 2341, 2346, 2348, 2349, 2363, 2364, 2366, 2373, 2382,
    2397, 2398, 2402, 2403, 2412, 2420, 2427, 2436, 2461, 2498, 2507, 2510, 2524,
    2530, 2535, 2537, 2570, 2597, 2609, 2610, 2616, 2619, 2624, 2646, 2650, 2660,
    2670, 2683, 2685, 2702, 2707, 2727, 2759, 2760, 2774, 2777, 2781, 2788, 2806,
    2807, 2810, 2829, 2840, 2882, 2888, 2890, 2895, 2902, 2908, 2909, 2910, 2922,
    2924, 2948, 2950, 2960, 2963, 2965, 2968, 2977, 3014, 3032, 3036, 3037, 3038,
    3043, 3063, 3081, 3088, 3091, 3122, 3128, 3158, 3179, 3181, 3192, 3193, 3200,
    3247, 3248, 3281, 3286, 3290, 3298, 3306, 3348, 3351, 3365, 3378, 3390, 3412,
    3420, 3426, 3459, 3460, 3461, 3473, 3496, 3507, 3518, 3529, 3535, 3538, 3545,
    3559, 3573, 3589, 3602, 3607, 3623, 3640, 3651, 3692, 3713, 3720, 3721, 3722,
    3752, 3753, 3775, 3777, 3778, 3794, 3808, 3831, 3832, 3860, 3868, 3869, 3877,
    3878, 3895, 3900, 3914, 3920, 3927, 3935, 3960, 3981, 3985, 3988, 3997, 4000,
    4001, 4002, 4004, 4006, 4038, 4064, 4071, 4078, 4079, 4085, 4089, 4090, 4101,
    4111, 4112, 4116, 4117, 4118, 4120, 4140, 4151, 4178, 4183, 4184, 4186, 4212,
    4272, 4277, 4287, 4330, 4340, 4346, 4349, 4357, 4364, 4405, 4442, 4459, 4460,
    4469, 4473, 4500, 4507, 4523, 4524, 4550, 4610, 4622, 4640, 4644, 4654, 4658,
    4659, 4672, 4682, 4683, 4684, 4685, 4691, 4756, 4776, 4779, 4784, 4785, 4801,
    4807, 4821, 4828, 4830, 4832, 4839, 4840, 4842, 4843, 4844, 4852, 4855, 4858,
    4865, 4870, 4894, 4895, 4900, 4906, 4911, 4959, 4968, 4969, 4970, 4972, 4983,
    4985, 4999, 5038, 5059, 5060, 5071, 5072, 5079, 5088. 5114, 5120, 5131, 5133,
    5134, 5135, 5148, 5177, 5178, 5182, 5183, 5211, 5227, 5235, 5267, 5270, 5275,
    5284, 5287, 5294, 5321, 5325, 5326, 5342, 5353, 5374, 5404, 5454, 5459, 5461,
    5466, 5474, 5492, 5494, 5495, 5532, 5554, 5555, 5566, 5576, 5577, 5578, 5584,
    5590, 5601, 5610, 5611, 5628, 5651, 5677, 5704, 5706, 5714, 5725, 5741, 5773,
    5794, 5799, 5801, 5827, 5843, 5853, 5868, 5884, 5897, 5904, 5906, 5956, 5985,
    5993, 6032, 6056, 6058, 6063, 6064, 6066, 6071, 6078, 6094, 6102, 6107, 6113,
    6158, 6162, 6170, 6177, 6179, 6184, 6201, 6221, 6224, 6229, 6237, 6238, 6246,
    6263, 6272, 6274, 6301, 6320, 6327, 6346, 6350, 6359, 6365, 6366, 6367, 6372,
    6374, 6380, 6387, 6398, 6406, 6423, 6428, 6440, 6452, 6453, 6454, 6456, 6457,
    6458, 6459, 6510, 6514, 6516, 6528, 6534, 6538, 6598

B

Beta gauge see Water saturation deficit, methods

Biliproteins see Salinity, ...; Water status in plant, effect on biliproteins

Bound water 12, 35, 81, 82, 83, 104, 135, 208, 218, 249, 269, 278, 285, 295, 296,
    403, 416, 420, 421, 422, 423, 424, 428, 441, 444, 459, 462, 473, 497, 521, 523,
    549, 550, 674, 683, 685, 686, 695, 696, 698, 704, 706, 741, 749, 750, 751, 768,
    772, 785, 794, 796, 819, 820, 853, 859, 860, 864, 899, 924, 999, 1046, 1127,
    1173, 1218, 1341, 1438, 1439, 1634, 1703, 1706, 1779, 1842, 1855, 1871, 1883,
    1892, 1935, 1963, 1991, 1997, 2142, 2147, 2352, 2378, 2430, 2471, 2491, 2492,
    2643, 2692, 2989, 2992, 2996, 3028, 3188, 3201, 3231, 3257, 3356, 3407, 3675,
    3688, 3711, 3866, 4201, 4202, 4203, 4229, 4310, 4387, 4388, 4397, 4412, 4441,
    4449, 4450, 4453, 4483, 4575, 4592, 4634, 4677, 4681, 4710, 4769, 4798, 4808,
    4810, 4812, 4817, 4890, 4936, 4945, 4948, 4949, 4988, 5036, 5061, 5075, 5094,
    5095, 5110, 5129, 5152, 5226, 5386, 5457, 5652, 5654, 5679, 5823, 5896, 6060,
    6067, 6087, 6088, 6105, 6128, 6146, 6206, 6211, 6213, 6214, 6275, 6296, 6342,
    6282, 6532, 6563

Bound water, methods 12, 35, 295, 341, 416, 420, 785, 859, 1842, 2146, 2491, 3702,
    4202, 4203, 4412, 4583, 5129

Boundary layer of air see Conductance for water vapour and CO_2 transfer, boundary
    layer of air

C

Canopy architecture see Drought, ...; Flooding, ...; Humidity of air, ...; Irriga-
    tion, ...; Precipitation and dew, ...; Soil moisture, ...; Water status in
    plant, effect on canopy architecture

Canopy model see Model of canopy

Carbohydrates, relation to conductance for water vapour and CO_2 transfer 5934

Carbohydrates, relation to stomata and epidermis 1476, 1513, 2163, 2197, 2475, 2817,
    3444, 4023, 5497, 6163, 6166

Carbohydrates, relation to water absorption by plant 3795, 6418

Carbohydrates, relation to water status in plant 1562, 1785, 3148, 4672, 4985, 5105,
    5237, 5255, 5886

Carbohydrates, relation to water transport in cells 2168, 2197, 6298

Carbohydrates, relation to wilting 1124, 1216, 2088, 2538, 2652, 3619, 4506, 4813,
    5931, 6298

Carbon fixation pathways see Drought, ...; Flooding, ...; Humidity of air, ...; Ir-
    rigation, ...; Osmotic agents, ...; Precipitation and dew, ...; Salinity, ...;
    Soil moisture, ...; Water status in plant, effect on carbon fixation pathways

Carbowax see Osmotic agents, ...

Carotenoids see Drought, ...; Irrigation, ...; Salinity, ...; Soil moisture, ...;
    Water status in plant, effect on carotenoids

Chlorophyll see Drought, ...; Flooding, ...; Humidity of air, ...; Irrigation, ...;
    Osmotic agents, ...; Precipitation and dew, ...; Salinity, ...; Soil moisture,
    ...; Water status in plant, effect on chlorophyll

Chloroplasts see Drought, ...; Humidity of air, ...; Osmotic agents, ...; Salinity,
    ...; Soil moisture, ...; Water status in plant, effect on chloroplasts

CO_2, effect on conductance for water vapour and CO_2 transfer 226, 306, 347, 376,
    616, 712, 875, 1180, 1387, 1617, 1715, 1876, 1944, 2063, 2422, 2658, 2751, 2773,
    2861, 2863, 3020, 3443, 3568, 3770, 3773, 3882, 4035, 4083, 4264, 4556, 4822,
    5200, 5311, 5335, 5435, 5574, 5674, 5934, 5958, 6103, 6120, 6321, 6340, 6417,
    6494, 6530

CO_2, effect on stomata and epidermis 84, 97, 98, 232, 306, 604, 709, 710, 711, 1056,
    1319, 1493, 1500, 1513, 1561, 1621, 1799, 1800, 1801, 1882, 1903, 1944, 2060,
    2063, 2065, 2111, 2327, 2673, 2677, 2731, 2801, 2942, 3170, 3255, 3327, 3439,
    3443, 3475, 3575, 3882, 4035, 4083, 5204, 5310, 5435, 5709, 5879, 5908, 5958,
    6041, 6103, 6123, 6163, 6164, 6166, 6225, 6226, 6321, 6478, 6479, 6494

CO_2, effect on transpiration 41, 97, 98, 616, 747, 1036, 1094, 1461, 1567, 1589,
    1715, 1720, 2673, 2975, 3020, 3093, 3665, 4556, 5262, 5381, 6093, 6110, 6576

CO_2, effect on water absorption by plant 1847, 3141, 3665

CO_2, effect on water transport in cells 6417

CO_2, effect on water transport in plant 1841, 4582

CO_2, effect on wilting 1074, 1368, 5918

CO_2 influx see Drought, ...; Flooding, ...: Humidity of air ...: Irrigation. ...;
    Osmotic agents, ...; Precipitation and dew, ...; Salinity, ...; Soil moisture,
    ...; Water status in plant, effect on CO_2 influx

Conductance for water vapour and CO_2 transfer, diurnal course 102, 143, 171, 198,
  199, 257, 279, 309, 333, 339, 340, 346, 362, 411, 472, 474, 478, 551, 568, 625,
  679, 766, 767, 870, 908, 1043, 1058, 1087, 1096, 1109, 1208, 1235, 1300, 1351,
  1363, 1370, 1493, 1506, 1642, 1653, 1701, 1702, 1808, 1851, 1915, 1916, 1944,
  1946, 2063, 2080, 2109, 2128, 2148, 2245, 2269, 2298, 2302, 2330, 2381, 2417,
  2467, 2482, 2501, 2545, 2558, 2619, 2677, 2682, 2691, 2734, 2735, 2737, 2778,
  2795, 2836, 2861, 2884, 2970, 2977, 2981, 3006, 3041, 3053, 3061, 3062, 3067,
  3107, 3131, 3157, 3165, 3205, 3261, 3287, 3349, 3350, 3354, 3392, 3405, 3417,
  3433, 3479, 3555, 3563, 3569, 3634, 3635, 3742, 3787, 3882, 3889, 3893, 3920,
  4025, 4042, 4051, 4137, 4214, 4215, 4266, 4267, 4306, 4307, 4324, 4353, 4380,
  4466, 4530, 4608, 4662, 4689, 4742, 4763, 4829, 4860, 4861, 4950, 5010, 5018,
  5054, 5081, 5105, 5152, 5167, 5180, 5341, 5350, 5406, 5435, 5463, 5486, 5546,
  5547, 5548, 5549, 5550, 5556, 5558, 5568, 5574, 5575, 5579, 5588, 5708, 5727,
  5744, 5753, 5783, 5854, 5870, 5871, 5895, 5910, 5922, 5928, 5965, 6027, 6031,
  6075, 6242, 6246, 6248, 6255, 6317, 6333, 6357, 6374, 6376, 6391, 6399, 6423,
  6506, 6519, 6526, 6533, 6559, 6587

Conductance for water vapour and CO_2 transfer, genetic 447, 730, 1036, 1531, 2105,
  2331, 2515, 3053, 3350, 3791, 3904, 4246, 5181, 5223, 5700, 5783, 6000, 6074,
  6189, 6240, 6586

Conductance for water vapour and CO_2 transfer, heterogeneity of single leaf blade
  1782, 3600, 4410, 5871

Conductance for water vapour and CO_2 transfer, intercellular spaces 151, 334, 590,
  810, 927, 1193, 1240, 1410, 1769, 2128, 3068, 3475, 3568, 3687, 3835, 3841,
  4156, 5278, 5310, 5335, 5381, 5768, 6225, 6486

Conductance for water vapour and CO_2 transfer, oscillations 226, 538, 1550, 2589,
  2862, 3061, 3107, 4635, 5470, 5548

Conductance for water vapour and CO_2 transfer, seasonal course 94, 102, 198, 257,
  282, 381, 568, 601, 679, 765, 788, 1356, 1363, 1370, 1701, 1803, 1915, 2064,
  2078, 2417, 2619, 2626, 2665, 2735, 2737, 2795, 2955, 3062, 3160, 3204, 3260,
  3585, 3635, 3691, 3742, 3774, 4051, 4157, 4214, 4267, 4307, 4608, 4767, 4974,
  5010, 5034, 5057, 5211, 5507, 5558, 5580, 5588, 5611, 5642, 5649, 5783, 5815,
  5905, 5910, 6026, 6027, 6118, 6255, 6317

Conductance for water vapour and CO_2 transfer, stomata 16, 20, 58, 69, 74, 75, 94,
  124, 134, 143, 151, 160, 161, 171, 183, 186, 188, 196, 198, 199, 201, 209, 222,
  226, 256, 279, 282, 304, 306, 308, 309, 313, 323, 333, 334, 339, 340, 347, 362,
  376, 379, 381, 384, 391, 397, 411, 415, 438, 447, 450, 451, 472, 474, 475, 486,
  487, 499, 527, 537, 538, 551, 552, 557, 568, 574, 589, 590, 592, 594, 596, 601,
  604, 606, 615, 625, 646, 656, 673, 679, 694, 712, 730, 737, 738, 741, 765, 766,
  781, 787, 788, 802, 808, 809, 810, 827, 828, 845, 861, 869, 870, 875, 897, 902,
  908, 911, 914, 917, 921, 927, 928, 933, 945, 955, 956, 968, 988, 993, 1031,
  1036, 1040, 1043, 1055, 1056, 1057, 1058, 1062, 1078, 1088, 1094, 1096, 1109,
  1126, 1134, 1140, 1155, 1159, 1169, 1193, 1195, 1203, 1209, 1235, 1240, 1280,
  1281, 1282, 1294, 1298, 1300, 1310, 1316, 1320, 1321, 1336, 1345, 1351, 1356,
  1370, 1378, 1387, 1397, 1416, 1431, 1444, 1447, 1461, 1468, 1469, 1492, 1500,
  1508, 1514, 1521, 1531, 1545, 1547, 1551, 1555, 1561, 1563, 1565, 1613, 1620,
  1621, 1623, 1640, 1641, 1642, 1645, 1653, 1654, 1655, 1670, 1686, 1702, 1714,
  1715, 1720, 1732, 1735, 1746, 1747, 1755, 1759, 1769, 1776, 1782, 1808, 1813,
  1832, 1845, 1856, 1860, 1867, 1876, 1887, 1903, 1915, 1920, 1944, 1946, 1948,
  1964, 1969, 2061, 2063, 2065, 2070, 2071, 2080, 2104, 2105, 2109, 2128, 2135,
  2185, 2203, 2213, 2216, 2217, 2229, 2243, 2298, 2302, 2320, 2328, 2331, 2332,
  2336, 2339, 2357, 2351, 2364, 2374, 2391, 2405, 2410, 2414, 2417, 2426, 2439,
  2452, 2454, 2468, 2481, 2485, 2487, 2501, 2515, 2529, 2531, 2541, 2545, 2551,
  2555, 2557, 2558, 2565, 2566, 2574, 2575, 2585, 2603, 2605, 2626, 2652, 2658,
  2665, 2677, 2691, 2698, 2700, 2701, 2719, 2734, 2737, 2751, 2761, 2766, 2787,
  2795, 2810, 2816, 2817, 2833, 2835, 2836, 2843, 2862, 2923, 2970, 2973, 2986,
  3020, 3021, 3040, 3041, 3047, 3048, 3053, 3067, 3068, 3100, 3106, 3111, 3131,
  3148, 3156, 3157, 3160, 3167, 3170, 3218, 3248, 3249, 3250, 3262, 3282, 3287,
  3293, 3294, 3323, 3351, 3354, 3392, 3393, 3396, 3397, 3405, 3417, 3433, 3437,

Conductance for water vapour and CO transfer, stomata (continued) 3443, 3451, 3474,
    3475, 3487, 3505, 3535, 3555, 3557, 3568, 3571, 3597, 3600, 3601, 3624, 3633,
    3635, 3657, 3660, 3687, 3691, 3770, 3774, 3779, 3783, 3791, 3806, 3822, 3835,
    3836, 3841, 3856, 3858, 3859, 3879, 3882, 3887, 3888, 3903, 3904, 3909, 3935,
    3952, 3957, 3970, 3971, 3974, 3975, 4024, 4025, 4041, 4051, 4060, 4065, 4089,
    4101, 4103, 4109, 4110, 4137, 4156, 4159, 4169, 4180, 4214, 4215, 4223, 4238,
    4264, 4267, 4287, 4302, 4324, 4331, 4342, 4347, 4355, 4380, 4392, 4407, 4408,
    4409, 4410, 4418, 4420, 4422, 4457, 4458, 4462, 4463, 4466, 4482, 4500, 4504,
    4505, 4511, 4512, 4556, 4600, 4601, 4604, 4609, 4631, 4663, 4673, 4676, 4684,
    4689, 4702, 4717, 4723, 4724, 4725, 4729, 4730, 4731, 4736, 4740, 4763, 4772,
    4773, 4774, 4778, 4822, 4829, 4832, 4860, 4861, 4863, 4864, 4882, 4893, 4926,
    4947, 4950, 4973, 4978, 5010, 5034, 5051, 5054, 5056, 5057, 5058, 5068, 5081,
    5087, 5103, 5105, 5139, 5152, 5163, 5181, 5184, 5186, 5189, 5197, 5210, 5211,
    5221, 5223, 5250, 5265, 5278, 5283, 5288, 5310, 5311, 5313, 5314, 5320, 5335,
    5337, 5338, 5339, 5341, 5350, 5372, 5380, 5381, 5383, 5385, 5395, 5411, 5412,
    5413, 5435, 5446, 5451, 5467, 5470, 5486, 5507, 5518, 5542, 5546, 5547, 5550,
    5553, 5557, 5574, 5575, 5580, 5594, 5596, 5600, 5602, 5603, 5631, 5677, 5682,
    5686, 5700, 5706, 5733, 5753, 5754, 5755, 5756, 5759, 5762, 5768, 5780, 5783,
    5807, 5815, 5840, 5841, 5854, 5862, 5870, 5871, 5879, 5881, 5895, 5905, 5910,
    5922, 5923, 5928, 5934, 5946, 5952, 5953, 5958, 5961, 5965, 5989, 6001, 6008,
    6025, 6026, 6031, 6041, 6059, 6074, 6075, 6117, 6120, 6130, 6131, 6132, 6153,
    6154, 6169, 6174, 6189, 6209, 6216, 6225, 6230, 6240, 6246, 6247, 6255, 6294,
    6295, 6297, 6313, 6321, 6332, 6333, 6334, 6339, 6340, 6357, 6374, 6376, 6378,
    6386, 6399, 6421, 6423, 6440, 6441, 6444, 6447, 6449, 6450, 6472, 6486, 6494,
    6495, 6506, 6527, 6529, 6530, 6533, 6559, 6575, 6577, 6586, 6587

Conductance for water vapour transfer, epidermis 20, 55, 183, 223, 305, 306, 307,
    312, 318, 338, 346, 379, 438, 522, 556, 604, 655, 658, 720, 738, 790, 886, 910,
    987, 1031, 1058, 1987, 1134, 1145, 1180, 1193, 1208, 1310, 1337, 1338, 1345,
    1362, 1363, 1370, 1409, 1416, 1431, 1468, 1493, 1503, 1506, 1523, 1530, 1617,
    1621, 1653, 1671, 1701, 1702, 1738, 1763, 1803, 1851, 1856, 1868, 1916,
    1964, 1990, 2128, 2148, 2204, 2208, 2215, 2245, 2255, 2269, 2276, 2320, 2330,
    2331, 2339, 2364, 2422, 2425, 2453, 2466, 2468, 2482, 2483, 2485, 2541, 2560,
    2582, 2583, 2589, 2614, 2619, 2631, 2632, 2657, 2682, 2697, 2700, 2735, 2766,
    2778, 2877, 2884, 2892, 2937, 2941, 2952, 2955, 2956, 2981, 3006, 3061, 3062,
    3072, 3081, 3107, 3109, 3110, 3131, 3165, 3204, 3206, 3219, 3252, 3260, 3261,
    3262, 3263, 3323, 3336, 3350, 3388, 3403, 3442, 3479, 3480, 3488, 3505, 3520,
    3556, 3563, 3564, 3569, 3588, 3606, 3620, 3624, 3634, 3667, 3669, 3689, 3739,
    3787, 3866, 3868, 3889, 3920, 3925, 3977, 3978, 3993, 4035, 4042, 4051, 4060,
    4083, 4089, 4157, 4159, 4179, 4224, 4246, 4266, 4267, 4274, 4306, 4307, 4324,
    4342, 4353, 4355, 4390, 4415, 4463, 4464, 4470, 4530, 4548, 4566, 4576, 4589,
    4594, 4606, 4608, 4635, 4660, 4662, 4663, 4664, 4690, 4701, 4717, 4722, 4742,
    4758, 4767, 4772, 4773, 4822, 4872, 4873, 4912, 4925, 4942, 4944, 4974, 4990,
    4996, 4997, 5018, 5051, 5056, 5058, 5167, 5180, 5181, 5200, 5210, 5286, 5304,
    5311, 5341, 5406, 5411, 5414, 5432, 5448, 5451, 5463, 5469, 5472, 5475, 5508,
    5514, 5515, 5540, 5547, 5548, 5549, 5551, 5558, 5568, 5573, 5579, 5588, 5611,
    5631, 5649, 5674, 5708, 5727, 5730, 5744, 5777, 5780, 5783, 5787, 5788, 5838,
    5842, 5849, 5868, 5880, 5881, 5909, 5915, 6000, 6015, 6027, 6032, 6044, 6103,
    6109, 6117, 6118, 6160, 6185, 6225, 6227, 6247, 6283, 6291, 6297, 6317, 6333,
    6391, 6392, 6394, 6403, 6415, 6423, 6430, 6450, 6464, 6468, 6508, 6519, 6526,
    6527

Consumption of water see Water consumption

Cryoscopy see Osmotic potential, methods

Cultivars, comparison during wilting 333, 445, 473, 899, 1016, 1017, 1108, 1622,
    1855, 2026, 2137, 2228, 2236, 2536, 2563, 3167, 3971, 4607, 4666, 4667, 4740,
    4888, 5109, 5132, 5252, 5413, 5783, 5825, 6317, 6437, 6495

Cultivars, comparison of conductance for water vapour and CO_2 transfer 308, 788,
    861, 1000, 1531, 2466, 2551, 2923, 2955, 2973, 3167, 3783, 3791, 3889, 4266,
    4306, 4702, 4740, 4763, 4863, 4978, 5180, 5210, 5211, 5286, 5372, 5406, 5413,
    5649, 5783, 5961, 6120, 6131, 6189, 6294, 6339, 6441, 6495, 6586

Drought, effect on water stress development 68, 593, 640, 1143, 1580, 2266, 2498, 2538, 3868, 4122, 4166, 4848, 4983, 5526, 5915

Drought, effect on water transport in cells 4872, 6518

Drought, effect on water transport in plant 269, 5654

Drought resistance 43, 58, 77, 91, 93, 136, 167, 183, 198, 208, 217, 218, 246, 250, 285, 287, 295, 333, 403, 417, 445, 447, 462, 464, 465, 466, 467, 470, 471, 477, 494, 514, 522, 634, 647, 662, 674, 676, 698, 751, 755, 756, 778, 794, 806, 813, 834, 836, 899, 926, 940, 954, 1016, 1026, 1030, 1060, 1074, 1107, 1108, 1144, 1162, 1200, 1240, 1269, 1290, 1308, 1318, 1386, 1395, 1418, 1420, 1456, 1503, 1573, 1577, 1580, 1689, 1706, 1709, 1710, 1736, 1737, 1775, 1798, 1900, 1935, 1963, 2009, 2088, 2091, 2095, 2100, 2103, 2137, 2219, 2228, 2249, 2273, 2288, 2325, 2335, 2339, 2346, 2348, 2350, 2358, 2363, 2380, 2384, 2385, 2445, 2482, 2493, 2501, 2504, 2519, 2536, 2542, 2545, 2555, 2580, 2605, 2623, 2652, 2674, 2681, 2738, 2790, 2798, 2804, 2866, 2885, 2893, 2927, 2954, 2956, 2988, 2999, 3011, 3012, 3025, 3069, 3121, 3122, 3161, 3183, 3189, 3211, 3276, 3351, 3362, 3381, 3417, 3554, 3603, 3675, 3730, 3734, 3744, 3749, 3789, 3799, 3815, 3819, 3820, 3846, 3847, 3848, 3893, 3894, 3963, 3964, 3978, 4011, 4013, 4014, 4018, 4027, 4094, 4100, 4144, 4145, 4146, 4147, 4148, 4177, 4192, 4218, 4236, 4250, 4270, 4289, 4293, 4333, 4348, 4369, 4371, 4408, 4409, 4426, 4449, 4450, 4451, 4456, 4510, 4562, 4574, 4607, 4627, 4666, 4667, 4671, 4674, 4677, 4692, 4745, 4748, 4749, 4759, 4788, 4792, 4820, 4848, 4862, 4867, 4888, 4899, 4913, 4946, 4964, 4965, 4979, 4988, 5061, 5127, 5132, 5133, 5189, 5195, 5204, 5235, 5238, 5251, 5304, 5324, 5373, 5376, 5384, 5450, 5467, 5476, 5478, 5594, 5622, 5623, 5626, 5628, 5632, 5633, 5634, 5635, 5676, 5727, 5745, 5775, 5783, 5784, 5791, 5807, 5825, 5840, 5842, 5849, 5879, 5891, 5915, 5935, 5965, 6025, 6074, 6083, 6120, 6122, 6131, 6139, 6144, 6186, 6208, 6209, 6215, 6240, 6259, 6261, 6293, 6349, 6402, 6411, 6422, 6428, 6436, 6441, 6480, 6495, 6497, 6504, 6505, 6525, 6568, 6592, 6593

Drought resistance, methods 2493, 3013, 3812, 3819, 4148, 4667

E

Ecotypes. comparison during wilting 17, 33, 1389, 1916, 2685, 3675, 4371

Ecotypes. comparison of conductance for water vapour and CO_2 transfer 188, 384, 679, 1410, 1620, 4060, 5507

Ecotypes, comparison of stomata and epidermis 47, 48, 160, 292, 384, 1010, 2555, 2744, 2802, 2841, 4060, 4504, 5224, 5507, 5630, 5636, 5743, 6180, 6321, 6322

Ecotypes, comparison of transpiration 17, 56, 160, 384, 765, 1037, 1187, 1234, 1862, 2293, 2555, 2879, 2986, 4446, 4544, 4638, 4696, 4963, 4988, 5297, 5506, 5555, 6427

Ecotypes, comparison of water absorption by plant 1309m 2180, 2427, 4752, 5555, 5904

Ecotypes, comparison of water status in plant 160, 199, 285, 335, 401, 679, 741, 1072, 1620, 1646, 1862, 2072, 2293, 2294, 2900, 4284, 4592, 4696, 4939, 4988, 5047, 5049, 5061, 6427, 6540

Ecotypes, comparison of water transport in plant 1910, 4940

Electron transport chain see Drought, ...; Flooding, ...; Humidity of air, ...; Osmotic agents, ...; Salinity, ...; Soil moisture, ...; Water status in plant, effect on electron transport chain

Enzyme inhibitors, effect on conductance for water vapour and CO_2 transfer 6577

Enzyme inhibitors, effect on stomata and epidermis 2552, 2991, 3255, 3317, 3444, 4525, 5497, 6165, 6321

Enzyme inhibitors, effect on transpiration 3317, 3628, 4345, 4954

Enzyme inhibitors, effect on water absorption by plant 3186, 3873, 4345

Enzyme inhibitors, effect on water status in plant 3112

Enzyme inhibitors, effect on water transport in cells 3186, 5220

Enzyme inhibitors, effect on wilting 3628, 3872

Enzymes, relation to stomata and epidermis 2054, 4437, 4695, 6033, 6200, 6321

Enzymes, relation to transpiration 2188, 2189

Enzymes, relation to water absorption by plant 1818, 2170, 6481

Enzymes, relation to water status in plant 1562, 2189, 2990, 3100, 4382, 5255, 5524, 5816, 5818

Enzymes, relation to wilting 1104, 1304, 1359, 1558, 1652, 2088, 3100, 5524, 5931, 6437, 6584

Epidermal conductance see Conductance for CO_2 transfer, epidermis; Conductance for water vapour transfer, epidermis

Epidermis see Abaxial and adaxial epidermes; Anatomical structure of epidermis; Conductance for CO_2 transfer, epidermis; Conductance for water vapour transfer, epidermis; Stomata and epidermis, ...; Stomata and epidermis, heterogeneity of single leaf blade

Evaporation 49, 53, 108, 144, 224, 327, 337, 360, 425, 468, 472, 504, 605, 654, 692, 716, 735, 738, 770, 789, 808, 816, 850, 866, 871, 954, 989, 1027, 1029, 1070, 1077, 1081, 1164, 1166, 1175, 1179, 1185, 1197, 1244, 1267, 1273, 1299, 1334, 1337, 1344, 1347, 1445, 1448, 1470, 1537, 1539, 1607, 1642, 1651, 1664, 1668, 1712, 1735, 1736, 1738, 1779, 1790, 1804, 1848, 1868, 1898, 1926, 1951, 1969, 2019, 2026, 2030, 2037, 2042, 2077, 2156, 2184, 2195, 2241, 2256, 2274, 2278, 2280, 2283, 2288, 2302, 2310, 2315, 2341, 2344, 2364, 2425, 2445, 2535, 2559, 2596, 2646, 2656, 2670, 2760, 2854, 2886, 2900, 2908, 2909, 2938, 2968, 3005, 3032, 3199, 3248, 3273, 3286, 3298, 3306, 3314, 3348, 3354, 3365, 3374, 3378, 3394, 3420, 3431, 3450, 3455, 3486, 3553, 3565, 3590, 3616, 3651, 3692, 3752, 3757, 3759, 3769, 3778, 3792, 3814, 3869, 3879, 3880, 3885, 3886, 3925, 3949, 3985, 3990, 4001, 4038, 4100, 4101, 4151, 4183, 4184, 4186, 4232, 4272, 4311, 4312, 4344, 4498, 4508, 4524, 4530, 4785, 4824, 4846, 4894, 4971, 4998, 5012, 5026, 5097, 5120, 5130, 5183, 5198, 5287, 5295, 5319, 5342, 5391, 5410, 5416, 5463, 5529, 5536, 5574, 5577, 5608, 5671, 5801, 5827, 5831, 5915, 5932, 6008, 6083, 6097, 6151, 6156, 6175, 6176, 6228, 6229, 6236, 6245, 6246, 6291, 6301, 6331, 6359, 6363, 6364, 6365, 6376, 6413, 6415, 6519, 6555

Evapotranspiration 49, 53, 93, 94, 100, 101, 105, 108, 114, 115, 144, 227, 242, 282, 312, 314, 315, 349, 350, 357, 360, 376, 383, 392, 405, 406, 411, 443, 452, 453, 454, 487, 507, 517, 554, 598, 625, 633, 634, 638, 654, 655, 656, 658, 668, 684, 714, 716, 717, 738, 758, 770, 789, 814, 816, 850, 856, 857, 866, 871, 872, 879, 910, 915, 925, 944, 956, 983, 989, 1003, 1012, 1070, 1074, 1102, 1103, 1116, 1123, 1140, 1142, 1165, 1166, 1184, 1187, 1196, 1197, 1201, 1238, 1263, 1267, 1273, 1278, 1299, 1318, 1334, 1335, 1360, 1370, 1372, 1381, 1394, 1398, 1448, 1491, 1499, 1501, 1519, 1522, 1524, 1529, 1533, 1581, 1595, 1616, 1642, 1651, 1672, 1710, 1729, 1738, 1747, 1753, 1754, 1760, 1761, 1762, 1788, 1789, 1815, 1822, 1884, 1889, 1915, 1963, 1968, 2002, 2037, 2063, 2080, 2093, 2103, 2109, 2156, 2165, 2195, 2224, 2233, 2234, 2253, 2261, 2263, 2279, 2280, 2288, 2296, 2310, 2314, 2341, 2342, 2343, 2344, 2381, 2440, 2498, 2504, 2516, 2522, 2529, 2559, 2616, 2619, 2632, 2638, 2646, 2647, 2653, 2660, 2687, 2724, 2729, 2760,

Evapotranspiration (continued) 2767, 2788, 2796, 2799, 2810, 2840, 2842, 2878, 2882,
    2890, 2908, 2914, 2924, 2938, 2946, 2950, 2964, 2965, 2977, 2978, 3005, 3014,
    3032. 3036, 3048, 3063, 3981, 3086, 3127, 3128, 3140, 3158, 3208, 3238, 3248,
    3266, 3296, 3333, 3341, 3348, 3361, 3363, 3365, 3382, 3393, 3394, 3472, 3517,
    3533, 3555, 3576, 3589, 3591, 3605, 3616, 3620, 3623, 3640, 3646, 3665, 3692,
    3707, 3713, 3764, 3778, 3781, 3784, 3785, 3792, 3808, 3821, 3831, 3850, 3880,
    3920, 3938, 3948, 3969, 3985, 3988, 4085, 4101, 4111, 4115, 4139, 4212, 4272,
    4283, 4320, 4330, 4340, 4344, 4364, 4388, 4399, 4454, 4468, 4497, 4500, 4508,
    4530, 4565, 4566, 4614, 4628, 4629, 4638, 4682, 4683, 4718, 4743, 4763, 4789,
    4790, 4837, 4859, 4918, 4950, 4970, 4992, 5012, 5026, 5037, 5038, 5078, 5080,
    5091, 5130, 5131, 5134, 5154, 5199, 5252, 5287, 5295, 5297, 5319, 5326, 5334,
    5342, 5367, 5380, 5414, 5416, 5420, 5437, 5454, 5462, 5463, 5474, 5501, 5505,
    5515, 5526, 5528, 5529, 5530, 5531, 5554, 5555, 5558, 5566, 5577, 5578, 5579,
    5593, 5611, 5625, 5697, 5706, 5794, 5801, 5830, 5831, 5868, 5869, 5932, 5941,
    5942, 5960, 6011, 6022, 6052, 6072, 6129, 6134, 6135, 6148, 6158, 6159, 6173,
    6174, 6175, 6176, 6211, 6237, 6246, 6264, 6272, 6283, 6291, 6320, 6327, 6353,
    6383, 6407, 6413, 6452, 6456, 6476, 6553, 6555, 6560

Evapotranspiration, methods, evaporimeters and lysimeters 49, 314, 360, 369, 596,
    871, 872, 925, 1082, 1267, 1287, 1335, 1353, 1448, 1519, 1607, 2077, 2151, 2195,
    2278, 2340, 2344, 2683, 2687, 2799, 2938, 2959, 2965, 2977, 3000, 3013, 3158,
    3238, 3243, 3363, 3417, 3647, 3707, 3949, 4038, 4311, 4312, 4419, 4454, 4790,
    4846, 5085, 5275, 5349, 5463, 5671, 5861, 5869, 6016, 6129, 6175, 6176, 6236,
    6555

Evapotranspiration, methods, other 49, 282, 319, 1123, 1179, 1263, 1267, 1273, 1448,
    1524, 2344, 2634, 2890, 2977, 2978, 3013, 3014, 3238, 3342, 3620, 3707, 5297,
    5463, 6175, 6176

F

Farming practices, effect on conductance for water vapour and CO_2 transfer 2213,
    3165, 5558, 6283

Farming practices, effect on stomata and epidermis 2213

Farming practices, effect on transpiration 507, 717, 989, 2037, 2505, 2647, 3165,
    3199, 3394, 3651, 4566, 5012, 5420, 5454, 5689, 6114, 6281, 6282, 6283, 6343,
    6383, 6413

Farming practices, effect on water absorption by plant 2004, 2005, 2032, 2084, 2112,
    2806, 2807, 2922, 2925, 3036, 3248, 3545, 3573, 4178, 4357, 4459, 5133, 5454,
    5584, 5610, 5985, 6102, 6201, 6452, 6453, 6456, 6457, 6458, 6538

Farming practices, effect on water status in plant 146, 169, 557, 1109, 2004, 4460,
    4936, 5454, 5558, 5903, 6067, 6275, 6281

Farming practices, effect on wilting 1392, 2873, 5133, 5252, 5454, 5686, 6526

Flooding, effect on canopy architecture 3574

Flooding, effect on carbon fixation pathways 5824

Flooding, effect on chlorophyll 2659, 2722, 3008, 4384, 4643, 4975, 5511, 5761, 6023

Flooding, effect on CO_2 influx 1251, 1566, 1909, 3008, 4643, 4889

Flooding, effect on conductance for water vapour and CO_2 transfer 3263, 4331, 5167,
    5881, 6120

Flooding, effect on electron transport chain 1999. 3246

Growth and productivity see D_2O, T_2O, ...; Drought, ...; Flooding, ...; Humidity of
     air, ...; Irrigation, ...; Osmotic agents, ...; Precipitation and dew, ...;
     Salinity, ...; Soil moisture, ...; Water status in plant, effect on growth and
     productivity

Growth substances, hormones, inhibitors, enzymes, antitranspirants etc., effect on
     stomatal aperture 84, 189, 448, 450, 529, 541, 576, 709, 710, 712, 773, 852,
     1126, 1282, 1513, 1640, 1833, 2327, 2361, 2607, 2991, 3633, 4169, 4546, 4791,
     5517, 6079, 6217, 6479

Growth substances, hormones, inhibitors etc., effect on conductance for water vapour
     and CO_2 transfer 58, 391, 450, 451, 712, 1180, 1714, 1944, 2063, 2064, 2405,
     2410, 2487, 2560, 2766, 2861, 3040, 3633, 3975, 3976, 4156, 4169, 4267, 4464,
     4673, 5139, 5934, 5946, 5961, 6209, 6448, 6449, 6525, 6577

Growth substances, hormones, inhibitors etc., effect on stomata and epidermis 58,
     140, 254, 303, 351, 361, 475, 498, 576, 628, 680, 709, 710, 711, 773, 1065,
     1130, 1248, 1274, 1275, 1320, 1513, 1561, 1758, 1798, 1799, 1800, 1827, 1944,
     1986, 2053, 2054, 2060, 2063, 2065, 2130, 2132, 2327, 2446, 2475, 2674, 2716,
     2731, 2766, 2817, 2831, 2850, 2942, 3011, 3040, 3060, 3114, 3131, 3275, 3291,
     3313, 3327, 3422, 3475, 3575, 3696, 3976, 4035, 4135, 4169, 4173, 4319, 4518,
     4526, 4546, 4673, 4699, 4739, 4907, 4908, 4914, 4991, 4997, 5112, 5123, 5139,
     5186, 5190, 5204, 5317, 5382, 5441, 5709, 5735, 5764, 5908, 5914, 5919, 5962,
     5984, 5986, 5987, 6025, 6050, 6073, 6165, 6204, 6217, 6225, 6226, 6232, 6244,
     6321, 6322, 6448, 6450, 6460, 6479, 6499, 6529, 6547, 6548, 6599

Growth substances, hormones, inhibitors etc., effect on transpiration 159, 161, 190,
     498, 516, 610, 712, 878, 1085, 1126, 1320, 1597, 1921, 2015, 2064, 2099, 2132,
     2225, 2307, 2325, 2361, 2369, 2374, 2549, 2550, 2607, 2766, 2942, 3026, 3027,
     3195, 3629, 3630, 3665, 3688, 3802, 3896, 3976, 3978, 4035, 4169, 4234, 4332,
     4399, 4416, 4424, 4472, 4546, 4642, 4668, 4760, 5111, 5186, 5226, 5317, 5961,
     5986, 6060, 6259, 6525

Growth substances, hormones, inhibitors etc., effect on water absorption by plant
     1085, 1597, 2132, 2225, 2834, 4513, 4760, 5716, 5834

Growth substances, hormones, inhibitors etc., effect on water status in plant 303,
     420, 503, 610, 858, 973, 1023, 1562, 1765, 1997, 2101, 2127, 2189, 2286, 2456,
     2487, 2560, 2627, 2628, 2726, 2831, 2993, 3131, 3312, 3527, 3629, 3630, 3633,
     3662, 3688, 3697, 3788, 3833, 3897, 3939, 3975, 4041, 4156, 4318, 4378, 4381,
     4429, 4585, 4653, 4688, 4755, 4835, 4949, 5014, 5039, 5055, 5204, 5226, 5560,
     5717, 5724, 6060, 6128, 6209

Growth substances, hormones, inhibitors etc., effect on water transport in cells 456,
     627, 2259, 3215, 3838, 4949, 5357, 6060, 6498

Growth substances, hormones, inhibitors etc., effect on water transport in plant 29,
     743, 1135, 1553, 1902, 2834, 4376, 4551, 4760, 5226, 5236, 5969

Growth substances, hormones, inhibitors etc., effect on wilting 58, 176, 322, 332,
     450, 498, 858, 995, 1124, 1130, 1304, 1737, 1829, 2100, 2369, 2406, 2449, 2487,
     2549, 2580, 2598, 2694, 2730, 2801, 2831, 3041, 3060, 3131, 3293, 3696, 3697,
     3788, 4029, 4087, 4131, 4169, 4172, 4210, 4271, 4318, 4369, 4584, 4632, 4762,
     4792, 4804, 4835, 4975, 5064, 5441, 5783, 5946, 6259, 6335, 6340, 6360, 6525,
     6580, 6599

Guard cells see Stomata, ...; Stomatal, ...

Guttation 443, 714, 1419, 1458, 1722, 2087, 2090, 2373, 3139, 3754, 4216, 4841,
     5214, 5244, 5267, 6136

H_2 isotopes see D_2O, T_2O, ...

Herbicides see Pesticides and herbicides, ...

Heterogeneity of single leaf blade see Conductance for water vapour and CO_2 transfer,
     ...; Stomata and epidermis, ...; Transpiration rate, ...; Water status in plant,
     heterogeneity of single leaf blade

Hormones see Growth substances, hormones, inhibitors etc., ...

Humidity of air, effect on canopy architecture 567, 988, 1761, 2233, 2251, 6326

Humidity of air, effect on carbon fixation pathways 5370, 5423, 6342

Humidity of air, effect on chlorophyll 13, 14, 1315, 3531, 4063, 5043, 5116, 5117,
     6082

Humidity of air, effect on chloroplasts 14, 88, 6581

Humidity of air, effect on CO_2 influx 65, 195, 209, 306, 307, 384, 400, 412, 477,
     478, 487, 537, 591, 765, 908, 909, 944, 988, 1461, 1550, 1638, 1646, 1670, 1671,
     1692, 2116, 2175, 2233, 2739, 2877, 3385, 3791, 3882, 3883, 4266, 4511, 4613,
     4885, 4963, 4997, 5056, 5087, 5304, 5322, 5479, 5545, 5811, 5899, 5909, 5910,
     5923, 5958, 6086, 6117, 6313, 6461, 6477

Humidity of air, effect on conductance for water vapour and CO_2 transfer 20, 143,
     226, 305, 306, 307, 339, 340, 376, 379, 478, 552, 601, 606, 766, 908, 1031, 1180,
     1235, 1321, 1363, 1370, 1493, 1550, 1613, 1620, 1653, 1654, 1670, 1747, 2128,
     2330, 2453, 2481, 2545, 2658, 2700, 2735, 2737, 2761, 2877, 2892, 2937, 3041,
     3067, 3072, 3107, 3260, 3261, 3405, 3433, 3437, 3441, 3442, 3474, 3479, 3505,
     3555, 3563, 3635, 3791, 3806, 3858, 3866, 3882, 4101, 4266, 4267, 4274, 4505,
     4511, 4742, 4861, 4872, 5056, 5057, 5058, 5087, 5152, 5283, 5337, 5341, 5343,
     5395, 5406, 5463, 5518, 5547, 5548, 5568, 5579, 5744, 5783, 5870, 5909, 5910,
     5923, 5958, 5965, 6008, 6103, 6117, 6130, 6131, 6248, 6297, 6313, 6321, 6403,
     6461, 6494, 6559

Humidity of air, effect on electron transport chain 882

Humidity of air, effect on growth and productivity 200, 251, 279, 330, 458, 472, 518,
     525, 552, 567, 617, 672, 814, 980, 997, 1258, 1313, 1332, 1340, 1460, 1499, 1638,
     1748, 1760, 1761, 1772, 1821, 1899, 1961, 2116, 2158, 2183, 2332, 2388, 2447,
     2453, 2459, 2668, 3132, 3330, 3351, 3532, 3659, 3750, 3804, 3807, 3866, 3883,
     3885, 4273, 4274, 4421, 4448, 4552, 4628, 4632, 4787, 4992, 5006, 5169, 5189,
     5489, 5518, 5587, 5644, 5707, 5767, 5833, 5836, 5990, 6007, 6036, 6048, 6142,
     6211, 6304, 6461, 6569, 6570

Humidity of air, effect on leaf anatomy 530, 708, 1832, 1839, 1899, 2433, 2655,
     3132, 3155, 3532, 3866, 3883, 4366, 4502, 5174, 5518, 6461, 6569

Humidity of air, effect on stomata and epidermis 84, 291, 306, 308, 351, 384, 475,
     601, 761, 767, 956, 1031, 1321, 1492, 1493, 1500, 1560, 1620, 1621, 1653, 1655,
     1830, 1832, 1934, 2067, 2175, 2332, 2351, 2446, 2481, 2555, 2574, 2737, 2761,
     2877, 2931, 3041, 3070, 3439, 3441, 3475, 3505, 3532, 3635, 3882, 4057, 4082,
     4194, 4267, 4313, 4317, 4466, 4504, 4505, 4515, 4516, 4611, 4684, 5010, 5547,
     5548, 5908, 5909, 5957, 5958, 5959, 5976, 5977, 6009, 6103, 6131, 6211, 6226,
     6321, 6461, 6494

Humidity of air, effect on transpiration 28, 40, 209, 233, 234, 242, 279, 307, 308,
     323, 379, 384, 392, 478, 487, 537, 545, 654, 655, 706, 758, 777, 809, 1031, 1084,
     1094, 1291, 1330, 1344, 1347, 1373, 1461, 1522, 1550, 1579, 1626, 1638, 1646,
     1671, 1788, 1692, 1788, 1934, 2187, 2215, 2233, 2263, 2293, 2422, 2511, 2555,
     2578, 2739, 2757, 2789, 2890, 2933, 3018, 3041, 3109, 3336, 3441, 3532, 3620,
     3712, 3787, 3791, 3832, 3878, 3914, 4082, 4273, 4274, 4314, 4345, 4421, 4595,
     4963, 4997, 5056, 5058, 5152, 5283, 5295, 5545, 5568, 5909, 5910, 5923, 6009,
     6136, 6167, 6211, 6297, 6313, 6461, 6537, 6555

L

Leaf insertion level, effect on transpiration 203, 255, 286, 544, 589, 610, 720,
     795, 827, 917, 918, 1209, 1934, 2071, 2375, 2454, 2485, 2636, 3067, 3188, 3601,
     3625, 4264, 4267, 4392, 4433, 4444, 4595, 4811, 5058, 5210, 6074, 6088, 6295

Leaf insertion level, effect on water status in plant 13, 139, 167, 199, 203, 356,
     376, 391, 403, 404, 470, 506, 542, 606, 661, 687, 688, 763, 910, 949, 1109, 1213,
     1324, 1338, 1399, 1562, 1577, 1611, 1650, 1845, 1851, 1871, 2025, 2250, 2330,
     2471, 2483, 2485, 2511, 2665, 2726, 2819, 2892, 2920, 2935, 2937, 2955, 3104,
     3188, 3376, 3653, 3669, 3679, 3711, 3774, 3921, 4026, 4041, 4064, 4217, 4268,
     4297, 4450, 4561, 4609, 4677, 4742, 4874, 4875, 4876, 4877, 4878, 4880, 5034,
     5069, 5073, 5074, 5106, 5164, 5235, 5318, 5332, 5336, 5654, 5708, 5728, 5729,
     5744, 5746, 5989, 6032, 6069, 6087, 6215, 6249, 6374, 6376, 6542

Leaf insertion level, effect on water transport in plant 4267, 5337

Leaf insertion level, effecton wilting 167, 687, 1622, 2071, 2454, 2538, 3279, 4217,
     4268, 4370, 4371, 4607, 4984, 5034, 5524

Leaf resistance see Conductance for water vapour and CO$_2$ transfer, ...

Leaf surface, waxes and trichomes 8, 117, 137, 264, 291, 628, 632, 672, 779, 996,
     1107, 1117, 1333, 1378, 1396, 1455, 1592, 1696, 1705, 1710, 1751, 1755, 1783,
     1852, 1901, 2078, 2081, 2167, 2193, 2324, 2459, 2519, 2753, 2783, 2784, 2785,
     2793, 2858, 2860, 2931, 3058, 3082, 3147, 3161, 3164, 3313, 3424, 3480, 3554,
     3734, 3789, 3816, 3930, 4060, 4077, 4105, 4106, 4134, 4188, 4248, 4255, 4289,
     4375, 4447, 4486, 4487, 4499, 4519, 4520, 4532, 4611, 4662, 4741, 5029, 5030,
     5044, 5098, 5155, 5258, 5293, 5363, 5365, 5493, 5539, 5540, 5553, 5615, 5742,
     5743, 5764, 5873, 5879, 5924, 5972, 6120, 6218, 6233, 6302, 6394

Leaf surface, waxes and trichomes, seasonal course 2078, 2623, 2795, 3585, 4204,
     5010, 5034,

Leaf temperature see Transpiration rate, effect of leaf temperature

Lipids and fatty acids, relation to stomata and epidermis 1366, 1462, 5186

Lipids and fatty acids, relation to transpiration 4738

Lipids and fatty acids, relation to water absorption by plant 1628, 4738

Lipids and fatty acids, relation to water status in plant 1202, 4176, 4738

Lipids and fatty acids, relation to wilting 2498, 2544, 3123, 5155

Lysimeters see Evapotranspiration, methods, evaporimeters and lysimeters

M

Manitol see Osmotic agents, ...

Mass flow see Water transport in plant, ...

Mathematical model see Model; Model of canopy

Matric potential in plant tissue (see also Water status in plant, ...) 4, 96, 475,
     522, 533, 687, 728, 784, 1270, 1620, 1698, 2430, 2441, 2503, 2545, 3103, 3655,
     3711, 5089, 5243, 5361, 5400, 5516, 5537, 6025, 6273

Matric potential in substrate 46, 49, 185, 337, 474, 536, 580, 668, 714, 722, 825,
     1072, 1278, 1337, 1428, 1655, 1681, 1695, 1802, 1804, 1877, 2001, 2002, 2069,
     2079, 2151, 2436, 2440, 2441, 2503, 2560, 2620, 2727, 2740, 2788, 2855, 2909,
     2910, 2964, 3012, 3095, 3125, 3393, 3431, 3518, 3655, 3806, 4103, 4115, 4189,
     4243, 5165, 5342, 5452, 5494, 5528, 5537, 5557, 5601, 5668, 5704, 5725, 5879,
     6016, 6229, 6237, 6323, 6324

Osmotic potential in plant tissue (continued) 2363, 2430, 2442, 2483, 2484, 2511,
    2522, 2541, 2545, 2546, 2552, 2576, 2619, 2626, 2627, 2628, 2665, 2676, 2681,
    2682, 2725, 2786, 2844, 2891, 2910, 2926, 2954, 3029, 3079, 3103, 3104, 3113,
    3145, 3148, 3162, 3166, 3167, 3175, 3222, 3228, 3229, 3231, 3284, 3292, 3293,
    3294, 3328, 3356, 3391, 3417, 3464, 3465, 3486, 3488, 3562, 3653, 3663, 3664,
    3677, 3680, 3689, 3704, 3751, 3774, 3779, 3838, 3863, 3891, 3897, 3927, 3939,
    3963, 3975, 4016, 4042, 4069, 4089, 4095, 4102, 4126, 4170, 4175, 4189, 4237,
    4266, 4267, 4273, 4274, 4301, 4327, 4328, 4335, 4342, 4348, 4356, 4369, 4371,
    4378, 4381, 4386, 4407, 4409, 4410, 4429, 4452, 4455, 4513, 4521, 4561, 4562,
    4563, 4588, 4590, 4609, 4618, 4619, 4620, 4621, 4638, 4669, 4679, 4690, 4732,
    4755, 4767, 4768, 4774, 4808, 4810, 4829, 4830, 4853, 4854, 4868, 4949, 5003,
    5007, 5031, 5032, 5034, 5039, 5047, 5049, 5089, 5103, 5105, 5106, 5107, 5110,
    5119, 5125, 5156, 5157, 5163, 5164, 5185, 5195, 5208, 5236, 5237, 5241, 5243,
    5249, 5290, 5304, 5318, 5320, 5321, 5350, 5361, 5395, 5400, 5411, 5415, 5447,
    5493, 5516, 5537, 5564, 5594, 5621, 5625, 5639, 5662, 5664, 5686, 5693, 5694,
    5709, 5730, 5740, 5762, 5770, 5781, 5786, 5788, 5809, 5816, 5840, 5841, 5842,
    5844, 5845, 5849, 5851, 5879, 5888, 5900, 5908, 5969, 5992, 6008, 6025, 6032,
    6062, 6069, 6081, 6144, 6215, 6222, 6321, 6322, 6332, 6333, 6334, 6391, 6416,
    6427, 6428, 6477, 6495, 6563, 6567, 6582, 6596

Osmotic potential in substrate 107, 154, 209, 214, 251, 346, 378, 379, 426, 503,
    509, 515, 546, 675, 676, 724, 978, 987, 1134, 1205, 1230, 1241, 1461, 1572, 1617,
    1678, 1811, 1824, 1874, 1928, 1950, 2043, 2069, 2118, 2202, 2229, 2286, 2405,
    2450, 2480, 2485, 2629, 2830, 2832, 2855, 2869, 2910, 2952, 2964, 2965, 2993,
    3091, 3176, 3417, 3527, 4154, 4176, 4189, 4217, 4273, 4309, 4669, 4747, 4820,
    5019, 5031, 5191, 5246, 5247, 5306, 5441, 5452, 5494, 5668, 5676, 5693, 5694,
    5714, 5770, 5782, 5818, 5865, 5982, 6009, 6070, 6084, 6199, 6265, 6323, 6324,
    6325, 6416, 6446, 6544, 6565,

Osmotic potential, methods 371, 666, 965, 984, 1713, 2069, 2095, 3013, 3029, 3592,
    4853, 5185, 5203

Oxygen, effect on conductance for water vapour and CO_2 transfer 1180, 1397, 3487,
    4482, 4664, 4731, 6399

Oxygen, effect on stomata and epidermis 450, 486, 500, 1513, 2673, 3487, 4525, 5262

Oxygen, effect on transpiration 828, 986, 2673, 2767, 3625, 5262, 5577, 5795

Oxygen, effect on water absorption by plant 99, 106, 546, 1847, 2480, 5166, 5852

Oxygen, effect on water status in plant 486, 546, 3487,

Oxygen, effect on water transport in cells 5220

Oxygen, effect on water transport in plant 99, 106, 546, 1910,

Oxygen, effect on wilting 486, 986, 1129, 1368

Ozone see Pollutants and ozone, ...

P

Pathological effects on conductance for water vapour and CO_2 transfer 186, 309, 945,
    1145, 1316, 1763, 1782, 2135, 2320, 2361, 3667, 5018, 5211, 5486, 6015

Pathological effects on stomata and epidermis 40, 222, 336, 395, 921, 1040, 1128,
    1231, 1316, 1551, 1560, 1645, 1693, 1763, 2067, 2135, 2351, 2361, 2428, 2458,
    2625, 3024, 3147, 3331, 4715, 5170, 5211, 5290, 5310, 5415, 6180, 6278, 6499,
    6583

Pathological effects on transpiration 40, 336, 527, 690, 691, 921, 945, 1040, 1252,
    1316, 1683, 1709, 1763, 1885, 2039, 2135, 2361, 2522, 2728, 3264, 3301, 3303,

Pathological effects on transpiration (continued) 3747, 4781, 4782, 4976, 5018, 5046, 5121, 5211, 5264, 5290, 5415, 5444, 5892, 5963, 6015, 6401

Pathological effects on water absorption by plant 981, 990, 1327, 1556, 1557, 1615, 1667, 1694, 1847, 1982, 2032, 2040, 2041, 2177, 2220, 2326, 2348, 2906, 3055, 3300, 3301, 3302, 3353, 3747, 4781, 4782, 4783, 5121, 5281, 5486, 5749, 6015, 6583

Pathological effects on water status in plant 186, 222, 336, 685, 920, 990, 1083, 1223, 1231, 1250, 1252, 1260, 1316, 1532, 1572, 1673, 1683, 1763, 1782, 1877, 1885, 1993, 2000, 2135, 2169, 2238, 2352, 2360, 2522, 2676, 2728, 2832, 2855, 3120, 3353, 4039, 4043, 4094, 4362, 4539, 4543, 4712, 4770, 4911, 5018, 5046, 5050, 5211, 5212, 5290, 5415, 5486, 5491, 5749, 5892, 6015, 6401

Pathological effects on water transport in plant 230, 309, 633, 732, 862, 921, 1073, 1316, 1407, 1782, 1834, 1861, 1907, 2108, 2135, 2319, 2470, 2540, 2728, 2852, 3303, 3353, 3667, 3747, 3771, 4299, 4976, 5018, 5486, 6251, 6252

Pathological effects on wilting 330, 697, 921, 947, 1250, 1260, 1297, 1558, 1709, 1859, 2122, 2131, 2160, 2190, 2275, 2320, 2326, 2413, 2517, 2522, 2608, 2661, 2675, 2828, 2906, 2915, 3303, 3352, 3399, 3747, 3972, 4030, 4040, 4492, 4545, 4593, 4781, 4911, 4976, 5046, 5121, 5213, 5214, 5281, 5282, 5290, 5491, 5749, 5819, 6251, 6252, 6466, 6583

Permanent wilting see Wilting, ...

Permeability see Water transport in cells, ...

Pesticides and herbicides, effect on conductance for water vapour and CO_2 transfer 945

Pesticides and herbicides, effect on stomata and epidermis 155, 1117, 1163, 1891, 1987, 2374, 6137, 6164, 6165, 6194, 6205, 6321

Pesticides and herbicides, effect on transpiration 155, 945, 1988, 2096, 2374, 2459, 3016, 3017, 3425, 4343, 5924, 6343, 6513

Pesticides and herbicides, effect on water absorption by plant 410, 1163, 2374, 3174, 4349, 4900, 5158, 5159, 6066, 6184

Pesticides and herbicides, effect on water status in plant 2076, 2144, 2145, 2621, 2819, 3797, 4262, 4276, 5011, 5312, 5441

Pesticides and herbicides, effect on water transport in cells 1038, 1879

Pesticides and herbicides, effect on water transport in plant 764, 1879, 1891, 2374

Pesticides and herbicides, effect on wilting 2460, 2517, 4770, 4771, 6404

pH, effect on conductance for water vapour and CO_2 transfer 6311

pH, effect on stomata and epidermis 653, 1132, 1513, 1755, 1800, 2690, 4518, 4693, 4694, 4695, 5497, 6232

pH, effect on transpiration 5137, 5874

pH, effect on water absorption by plant 581, 2316, 5055, 6221

pH, effect on water transport in cells 2166, 2167, 2168, 3215, 3334, 4373, 5220, 6311, 6585

pH, effect on water transport in plant 5521

pH, effect on wilting 1308, 2219,

Precipitation and dew, effect on growth and productivity (continued) 2912, 2914,
    2967, 3038, 3091, 3118, 3198, 3225, 3316, 3319, 3326, 3333, 3338, 3394, 3395,
    3404, 3476, 3501, 3545, 3604, 3623, 3637, 3650, 3690, 3729, 3730, 3756, 3804,
    3807, 3845, 3870, 3926, 3988, 4001, 4058, 4111, 4120, 4174, 4206, 4277, 4430,
    4489, 4665, 4682, 4799, 4821, 4834, 4857, 4884, 4919, 4921, 4922, 4959, 4980,
    5070, 5114, 5132, 5234, 5356, 5367, 5431, 5440, 5488, 5490, 5559, 5563, 5584,
    5587, 5606, 5610, 5667, 5683, 5711, 5715, 5752, 5791, 5796, 5884, 5933, 5980,
    5985, 6020, 6077, 6094, 6115, 6188, 6202, 6241, 6268, 6351, 6352, 6353, 6362,
    6366, 6372, 6373, 6420, 6432, 6451, 6455

Precipitation and dew, effect on leaf anatomy 1839, 2854, 4153, 4205, 5356

Precipitation and dew, effect on respiration 6490

Precipitation and dew, effect on stomata and epidermis 3554, 6118

Precipitation and dew, effect on transpiration 312, 468, 738, 866, 1342, 1538, 1579,
    1651, 1889, 2184, 2195, 2296, 2578, 3589, 4139, 5454, 5869, 6490

Precipitation and dew, effect on water absorption by plant 649, 1312, 1663, 1694,
    1743, 1982, 2984, 2112, 2121, 2246, 2760, 3121, 3378, 4112, 4818, 4821, 5114,
    6359

Precipitation and dew, effect on water status in plant 223, 285, 397, 400, 432, 436,
    557, 1203, 1730, 2128, 2144, 2145, 2339, 2943, 3876, 4073, 4296, 4936, 5806

Precipitation and dew, effect on water transport in plant 1211, 1293, 2165, 3257

Precipitation and dew, effect on wilting 317, 738, 1077, 1343, 1389, 1969, 2214,
    2742, 4112,

Pressure bomb see Water potential, methods, pressure bomb

Pressure potential in plant tissue (see also Water status in plant, ...) 4, 72, 77,
    109, 129, 139, 147, 152, 153, 174, 250, 276, 332, 352, 376, 378, 401, 404, 406,
    470, 471, 475, 494, 522, 540, 556, 613, 614, 677, 687, 688, 728, 733, 745, 783,
    784, 787, 810, 833, 834, 877, 908, 910, 976, 977, 987, 1010, 1030, 1054, 1074,
    1096, 1098, 1155, 1168, 1185, 1226, 1227, 1246, 1247, 1255, 1264, 1270, 1296,
    1329, 1349, 1367, 1376, 1505, 1526, 1562, 1563, 1620, 1654, 1655, 1669, 1683,
    1736, 1740, 1832, 1841, 1845, 1928, 1929, 1930, 1945, 2017, 2028, 2031, 2095,
    2149, 2154, 2171, 2210, 2215, 2229, 2330, 2362, 2478, 2484, 2485, 2531, 2541,
    2545, 2546, 2604, 2652, 2671, 2681, 2682, 2731, 2761, 2764, 2778, 2816, 2844,
    2847, 2869, 2891, 2892, 2926, 2954, 3029, 3048, 3061, 3068, 3103, 3104, 3145,
    3148, 3222, 3231, 3233, 3282, 3293, 3334, 3556, 3376, 3391, 3435, 3436, 3488,
    3502, 3509, 3527, 3680, 3681, 3689, 3709, 3744, 3774, 3793, 3838, 3866, 3891,
    3963, 3964, 3972, 3975, 3978, 4089, 4098, 4099, 4102, 4126, 4170, 4189, 4267,
    4273, 4274, 4284, 4294, 4301, 4342, 4348, 4356, 4371, 4380, 4407, 4408, 4409,
    4410, 4449, 4452, 4481, 4485, 4561, 4562, 4588, 4617, 4662, 4679, 4755, 4767,
    4774, 4854, 4928, 5003, 5015, 5020, 5055, 5089, 5103, 5105, 5107, 5163, 5164,
    5203, 5241, 5242, 5243, 5249, 5254, 5304, 5318, 5333, 5361, 5395, 5400, 5411,
    5516, 5537, 5594, 5625, 5639, 5656, 5664, 5682, 5686, 5709, 5712, 5730, 5739,
    5740, 5745, 5786, 5788, 5816, 5840, 5844, 5845, 5849, 5879, 5908, 5919, 5929,
    5946, 5962, 5969, 5977, 6010, 6025, 6026, 6032, 6069, 6257, 6303, 6313, 6321,
    6323, 6332, 6333, 6334, 6495, 6563, 6596, 6602

Pressure potential, methods 150, 332, 540, 833, 834, 877, 976, 1713, 2017, 2095,
    4294, 5003, 5241, 5537, 5712

Productivity see Growth and productivity ...

Productivity of algae see Osmotic agents, ...; Salinity, effect on productivity of
    algae

Productivity of transpiration 46, 49, 63, 65, 116, 125, 134, 142, 172, 188, 198,
    201, 206, 209, 221, 246, 258, 307, 318, 346, 370, 385, 463, 534, 544, 579, 593,
    616, 684, 714, 765, 771, 802, 804, 810, 812, 822, 869, 870, 876, 884, 904, 917,
    918, 1074, 1094, 1097, 1190, 1234, 1269, 1319, 1334, 1335, 1387, 1461, 1493,
    1500, 1508, 1538, 1567, 1568, 1621, 1626, 1651, 1670, 1671, 1684, 1686, 1690,
    1692, 1710, 1720, 1798, 1815, 1916, 1934, 1969, 1970, 1978, 1995, 2057, 2065,
    2071, 2105, 2214, 2288, 2454, 2485, 2529, 2533, 2537, 2538, 2545, 2550, 2563,
    2573, 2586, 2646, 2677, 2683, 2686, 2740, 2863, 2884, 2923, 2983, 3036, 3052,
    3067, 3089, 3092, 3093, 3102, 3199, 3205, 3241, 3251, 3252, 3282, 3305, 3311,
    3336, 3351, 3375, 3480, 3489, 3502, 3548, 3549, 3565, 3571, 3572, 3600, 3601,
    3623, 3651, 3678, 3691, 3715, 3717, 3850, 3855, 3856, 3909, 3970, 3992, 3993,
    3994, 4001, 4035, 4101, 4157, 4158, 4214, 4239, 4392, 4508, 4595, 4662, 4671,
    4685, 4801, 4829, 4849, 4925, 4963, 4985, 4986, 5088, 5201, 5261, 5325, 5348,
    5350, 5352, 5372, 5381, 5395, 5399, 5424, 5458, 5461, 5463, 5494, 5538, 5541,
    5544, 5545, 5593, 5649, 5678, 5686, 5718, 5719, 5783, 5788, 5829, 5855, 5888,
    5901, 5908, 5947, 6005, 6032, 6056, 6117, 6131, 6132, 6151, 6159, 6179, 6189,
    6225, 6226, 6230, 6264, 6271, 6334, 6339, 6352, 6367, 6378, 6441, 6456, 6476,
    6515, 6576

Proteins, amino acids, nucleic acids, relation to stomata and epidermis 1476

Proteins, amino acids, nucleic acids, relation to water absorption by plant 1079,
    1341, 2046, 2047, 2139, 2232

Proteins, amino acids, nucleic acids, relation to water status in plant 1054, 1202,
    1239, 1562, 2189, 4217, 4382, 5920, 6393

Proteins, amino acids, nucleic acids, relation to water transport in cells 4373

Proteins, amino acids, nucleic acids, relation to water transport in plant 1311,

Proteins, amino acids, nucleic acids, relation to wilting 1104, 1108, 1124, 1216,
    1304, 1900, 2088, 2621, 2652, 2927, 3514, 3609, 3619, 4219, 4541, 4759, 4835,
    4836, 5101, 5324, 5888, 6186

Psychrometry see Water potential, methods, psychrometry

R

Radiation see Irradiance, ...

Rain see Precipitation and dew, ...

Reactivity of stomata see Stomatal reaction rate; Stomatal reactivity during leaf
    ontogenesis

Rehydratation 58, 63, 66, 67, 69, 135, 156, 167, 188, 310, 318, 340, 381, 401, 435,
    446, 450, 460, 461, 474, 506, 571, 640, 661, 697, 757, 781, 790, 794, 810, 870,
    895, 906, 907, 940, 960, 995, 1030, 1043, 1074, 1108, 1125, 1129, 1130, 1133,
    1169, 1216, 1229, 1240, 1259, 1260, 1272, 1290, 1303, 1358, 1359, 1395, 1396,
    1423, 1507, 1547, 1597, 1613, 1638, 1709, 1763, 1786, 1855, 1900, 1916, 1935,
    1940, 2149, 2205, 2243, 2273, 2346, 2358, 2390, 2419, 2432, 2472, 2487, 2494,
    2501, 2545, 2546, 2549, 2560, 2564, 2567, 2583, 2655, 2679, 2682, 2684, 2695,
    2696, 2698, 2710, 2716, 2801, 2804, 2820, 2866, 2867, 2923, 2957, 2979, 2990,
    3075, 3103, 3109, 3130, 3142, 3152, 3153, 3278, 3358, 3374, 3388, 3451, 3464,
    3465, 3505, 3642, 3671, 3698, 3710, 3712, 3788, 3799, 3800, 3809, 3866, 3874,
    3962, 3974, 3999, 4013, 4014, 4018, 4065, 4103, 4131, 4150, 4152, 4293, 4295,
    4320, 4353, 4420, 4431, 4464, 4534, 4541, 4546, 4574, 4577, 4578, 4594, 4636,
    4661, 4664, 4670, 4678, 4724, 4749, 4762, 4818, 4829, 4964, 4997, 5025, 5097,
    5102, 5136, 5204, 5308, 5324, 5373, 5378, 5384, 5388, 5413, 5430, 5487, 5542,
    5543, 5567, 5623, 5632, 5687, 5688, 5728, 5760, 5783, 5788, 5891, 5910, 5916,
    5946, 5962, 6025, 6034, 6132, 6276, 6279, 6321, 6331, 6469

Relative water content see Water saturation deficit

Replica method see Stomatal aperture, methods, microrelief methods

Resistance see Conductance ...

Respiration see Drought, ...; Flooding, ...; Irrigation, ...; Osmotic agents, ...;
    Precipitation and dew, ...; Salinity, ...; Soil moisture, ...; Water status in
    plant, effect on respiration

Root pressure, exudation 29, 30, 99, 237, 239, 321, 343, 561, 577, 581, 922, 942,
    1019, 1163, 1173, 1232, 1249, 1553, 1585, 1587, 1597, 1653, 1814, 1866, 1870,
    1874, 1910, 1969, 2087, 2090, 2222, 2254, 2305, 2365, 2581, 2763, 2779, 2834,
    2880, 3141, 3145, 3343, 3353, 3645, 3795, 3838, 3863, 4216, 4618, 4619, 4620,
    4621, 4669, 4746, 4760, 4825, 5236, 5362, 5716, 5828, 5919, 5969, 5992, 6015,
    6025, 6062

Root pressure, exudation, methods 1268, 2370,

S

Saccharides see Carbohydrates, ...

Saline water see Irrigation water quality

Salinity, effect on biliproteins 5260

Salinity, effect on carbon fixation pathways 469, 676, 1076, 1568, 1631, 1886, 1912,
    2449, 2962, 3078, 3569, 4107, 4108, 4215, 4359, 4553, 4726, 4764, 4904, 5156,
    5208, 5350, 5369, 5499, 5586, 5737, 6231, 6484

Salinity, effect on carotenoids 5260

Salinity, effect on chlorophyll 943, 1023, 1631, 1766, 1795, 2123, 2384, 2740, 2962,
    3068, 3470, 3611, 3833, 3983, 4244, 4361, 4726, 5260, 5604, 6504, 6505, 6507

Salinity, effect on chloroplasts 742, 890, 2384, 4623, 6222

Salinity, effect on CO_2 influx 3, 52, 209, 250, 251, 676, 1527, 1629, 1678, 1685,
    1766, 1795, 2449, 2641, 2719, 2952, 2987, 3068, 3408, 3569, 3687, 3854, 3896,
    3983, 3992, 3994, 4215, 4342, 4360, 4514, 4553, 4624, 4672, 4719, 4726, 4904,
    5014, 5156, 5260, 5298, 5586, 5737, 5757, 5784, 5887, 5952, 6527

Salinity, effect on conductance for water vapour and CO_2 transfer 250, 676, 987,
    1500, 2952, 3569, 3992, 3994, 4215, 4274, 4342, 4514, 5167, 5411, 5952, 6216,
    6527

Salinity, effect on electron transport chain 469, 676, 1912, 4553, 4904, 5737

Salinity, effect on growth and productivity 3, 64, 89, 119, 178, 210, 251, 346, 378,
    394, 617, 629, 650, 675, 676, 771, 793, 801, 846, 854, 855, 943, 986, 987, 1039,
    1047, 1082, 1136, 1220, 1226, 1241, 1277, 1404, 1460, 1486, 1562, 1570, 1593,
    1639, 1684, 1766, 1770, 1794, 1843, 1844, 1931, 1955, 1962, 2085, 2099, 2123,
    2129, 2159, 2194, 2221, 2305, 2489, 2522, 2530, 2629, 2719, 2725, 2738, 2740,
    2769, 2775, 2786, 2882, 2887, 2891, 2949, 2971, 3068, 3084, 3085, 3091, 3101,
    3184, 3239, 3244, 3265, 3269, 3285, 3357, 3418, 3423, 3517, 3594, 3612, 3613,
    3668, 3714, 3733, 3746, 3796, 3803, 3813, 3830, 3833, 3842, 3846, 3852, 3853,
    3854, 3902, 3991, 3994, 4015, 4017, 4031, 4032, 4033, 4069, 4116, 4123, 4124,
    4160, 4244, 4273, 4274, 4288, 4325, 4364, 4406, 4514, 4651, 4672, 4682, 4719,
    4720, 4867, 4871, 4929, 4932, 5001, 5014, 5031, 5032, 5108, 5126, 5207, 5255,
    5260, 5303, 5321, 5330, 5331, 5359, 5360, 5369, 5382, 5456, 5496, 5503, 5516,
    5583, 5584, 5586, 5612, 5645, 5650, 5661, 5663, 5720, 5722, 5789, 5835, 5912,

Salinity, effect on growth and productivity (continued) 5952, 6017, 6084, 6096,
    6097, 6111, 6162, 6182, 6183, 6198, 6199, 6222, 6234, 6235, 6324, 6325, 6348,
    6395, 6396, 6435, 6436, 6455, 6500, 6504, 6505, 6507, 6523, 6527, 6539, 6544,
    6565

Salinity, effect on leaf anatomy 676, 1631, 2099, 4514, 4672, 4897, 4929, 5259,
    5329, 5952, 6308, 6544

Salinity, effect on photorespiration 2987, 4672, 4726, 5369, 5419

Salinity, effect on productivity of algae 1527, 4361

Salinity, effect on respiration 1637, 1678, 2952, 3408, 3687, 4342, 4625, 4672,
    4719, 6042, 6518

Salinity, effect on stomata and epidermis 250, 271, 675, 676, 996, 1130, 1136, 2304,
    2305, 2485, 4726, 4897, 4929, 5259, 6308

Salinity, effect on transpiration 209, 250, 346, 475, 676, 771, 987, 1039, 1136,
    1220, 1350, 1684, 1685, 1862, 2099, 2305, 2449, 2740, 2882, 3068, 3517, 3537,
    3569, 3687, 3796, 3896, 3992, 3994, 4215, 4274, 4342, 4364, 4672, 5110, 5757,
    6306, 6307, 6527

Salinity, effect on water absorption by plant 103, 1082, 1634, 1950, 2124, 2305,
    2480, 2775, 2887, 2964, 3085, 3269, 3295, 3357, 3421, 3668, 3803, 3902, 4124,
    4160, 4672, 5165, 5166, 5167, 5494

Salinity, effect on water status in plant 72, 210, 249, 250, 617, 771, 949, 1023,
    1076, 1255, 1324, 1439, 1634, 1637, 1862, 1962, 2092, 2123, 2305, 2725, 2869,
    2891, 3593, 3689, 3704, 3796, 3833, 3851, 3991, 3994, 4154, 4249, 4273, 4274,
    4342, 4672, 4786, 5110, 5208, 5255, 5259, 5333, 5411, 5452, 5607, 5693, 5770.
    5845, 6192, 6216, 6222, 6306, 6307, 6416, 6436, 6527, 6539, 6540

Salinity, effect on water transport in cells 4113, 5329, 6172, 6518, 6585

Salinity, effect on water transport in plant 250, 269, 776, 1910, 4053

Salinity, effect on wilting 250, 986, 1039, 1129, 1130, 2478, 2629, 3148, 3848, 4124,
    4166, 4274, 4286, 4938, 5031, 5206, 5320, 5693, 5694, 5770

Seasonal course see Conductance for water vapour and CO_2 transfer, ...; Leaf surface,
    waxes and trichomes, ...; Stomatal aperture, ...; Transpiration rate, ...; Water
    absorption by plant, ...; Water status in plant, ...; Water transport in plant,
    ,..; Wilting, seasonal course

Senescence see Age of plant, ...

Sensible heat transfer see Transpiration rate, theoretical background

Shrinkage of plant see Water content in plant and related volume changes

Simulation see Model; Model of canopy

Soil moisture, effect on canopy architecture 567, 973, 1523, 4973

Soil moisture, effect on carbon fixation pathways 622, 1946, 2483, 3569, 4215, 4643
    5350, 5352, 5499, 5581

Soil moisture, effect on carotenoids 1341, 1963, 2055, 3003, 4642, 5272

Soil moisture, effect on chlorophyll 13, 92, 462, 588, 634, 796, 806, 1192, 1341,
    1525, 1547, 1963, 2055, 2402, 2822, 2911, 3087, 3272, 3761, 4152, 4396, 4642,
    4643, 4899, 4952, 4985, 5033, 5171, 5272, 5631, 5798, 6504, 6505

Soil moisture, effect on growth and productivity (continued) 5266, 5270, 5276, 5285,
    5290, 5292, 5367, 5374, 5375, 5393, 5398, 5421, 5426, 5453, 5461, 5465, 5473,
    5496, 5498, 5505, 5512, 5513, 5516, 5518, 5530, 5531, 5532, 5536, 5557, 5562,
    5575, 5576, 5577, 5587, 5591, 5593, 5596, 5597, 5598, 5611, 5633, 5640, 5651,
    5653, 5668, 5704, 5713, 5714, 5719, 5721, 5725, 5726, 5763, 5778, 5791, 5805,
    5807, 5821, 5831, 5847, 5853, 5856, 5872, 5874, 5876, 5879, 5888, 5893, 5908,
    5911, 5918, 5921, 5947, 5956, 5982, 5999, 6006, 6007, 6011, 6014, 6018, 6035,
    6036, 6039, 6044, 6048, 6051, 6061, 6064, 6065, 6071, 6077, 6078, 6096, 6097,
    6116, 6126, 6135, 6140, 6145, 6147, 6148, 6151, 6158, 6159, 6162, 6170, 6183,
    6212, 6214, 6220, 6224, 6238, 6269, 6272, 6289, 6293, 6300, 6304, 6320, 6324,
    6333, 6344, 6348, 6350, 6362, 6363, 6366, 6367, 6368, 6369, 6374, 6380, 6398,
    6406, 6411, 6432, 6440, 6445, 6447, 6470, 6476, 6480, 6483, 6504, 6505, 6514,
    6526, 6535, 6546, 6551, 6560, 6564, 6569, 6570, 6571, 6572, 6592, 6593, 6594,
    6601

Soil moisture, effect on leaf anatomy 80, 324, 338, 366, 491, 634, 959, 1053, 1839,
    2682, 2892, 3155, 3206, 3535, 3603, 3674, 3723, 3734, 3824, 3930, 4008, 4199,
    4459, 4502, 4658, 4672, 4690, 4888, 4961, 4973, 4974, 4990, 5033, 5169, 5289,
    5398, 5518, 5532, 5557, 5575, 5581, 5653, 5807, 5874, 6054, 6092, 6100, 6293,
    6398, 6423, 6569

Soil moisture, effect on photorespiration 2863, 4528, 4672

Soil moisture, effect on respiration 288, 522, 599, 610, 634, 751, 1963, 2592, 2739,
    2835, 2863, 3736, 4012, 4391, 4448, 4528, 4672, 4934, 5191, 5485, 5493, 5668,
    6145, 6207

Soil moisture, effect on stomata and epidermis 136, 216, 338, 381, 384, 576, 956,
    1162, 1555, 1710, 2067, 2302, 2481, 2555, 2787, 2810, 3235, 3313, 3505, 3532,
    3589, 3603, 3734, 3893, 3904, 3930, 4089, 4194, 4210, 4347, 4420, 4504, 4546,
    4643, 4671, 4798, 5105, 6009, 6041, 6092, 6100, 6321

Soil moisture, effect on transpiration 69, 198, 213, 247, 261, 263, 327, 357, 381,
    383, 384, 453, 459, 487, 491, 511, 579, 597, 610, 616, 654, 706, 738, 764, 789,
    810, 813, 817, 850, 879, 910, 1037, 1053, 1196, 1252, 1308, 1381, 1416, 1424,
    1437, 1501, 1653, 1815, 1963, 2037, 2240, 2307, 2481, 2490, 2504, 2542, 2545,
    2555, 2578, 2687, 2739, 2808, 2863, 2933, 2977, 3003, 3041, 3048, 3179, 3417,
    3505, 3532, 3535, 3569, 3572, 3589, 3603, 3734, 3781, 3806, 3878, 3904, 3914,
    4038, 4071, 4101, 4150, 4215, 4283, 4320, 4421, 4528, 4565, 4636, 4642, 4643,
    4671, 4796, 4797, 4829, 4858, 4895, 4899, 4963, 5058, 5148, 5195, 5244, 5261,
    5290, 5342, 5380, 5458, 5463, 5467, 5494, 5505, 5532, 5874, 5910, 6009, 6026,
    6134, 6148, 6214, 6246, 6259, 6340, 6468

Soil moisture, effect on water absorption by plant 146, 453, 561, 871, 883, 1039,
    1312, 1341, 2198, 2872, 2902, 2903, 3095, 3426, 3778, 4752, 4842, 4843, 5183,
    5244, 5389, 5579, 5696, 5983, 6148, 6331, 6468

Soil moisture, effect on water status in plant 13, 26, 61, 63, 106, 133, 167, 193,
    198, 208, 213, 263, 335, 338, 339, 358, 378, 446, 459, 462, 474, 487, 522, 526,
    552, 580, 588, 606, 664, 704, 706, 722, 749, 820, 870, 899, 910, 1013, 1030,
    1048, 1072, 1138, 1142, 1143, 1252, 1266, 1299, 1324, 1341, 1349, 1352, 1437,
    1439, 1541, 1547, 1622, 1638, 1728, 1747, 1786, 1802, 1804, 1838, 1870, 1871,
    1990, 1922, 2072, 2079, 2117, 2204, 2236, 2242, 2250, 2276, 2328, 2440, 2462,
    2483, 2484, 2510, 2511, 2545, 2605, 2619, 2627, 2643, 2682, 2808, 2863, 2888,
    2892, 2933, 2934, 2935, 2957, 3003, 3226, 3417, 3532, 3544, 3563, 3564, 3593,
    3642, 3676, 3689, 3710, 3734, 3779, 3815, 3864, 3914, 3927, 3981, 4045, 4062,
    4064, 4071, 4194, 4214, 4239, 4321, 4356, 4408, 4528, 4634, 4643, 4670, 4672,
    4684, 4690, 4762, 4792, 4895, 4949, 4974, 4990, 5073, 5074, 5148, 5195, 5388,
    5395, 5467, 5468, 5518, 5550, 5557, 5575, 5788, 5822, 5874, 6018, 6032, 6044,
    6132, 6143, 6156, 6214, 6323, 6332, 6376, 6468, 6568, 6569, 6592

Soil moisture, effect on water transport in cells 2510

Stomatal aperture (continued)   524, 576, 601, 648, 673, 679, 710, 788, 808, 852, 920,
    921, 928, 956, 1031, 1040, 1057, 1058, 1078, 1088, 1128, 1130, 1134, 1159, 1195,
    1203, 1280, 1294, 1319, 1320, 1336, 1513, 1555, 1559, 1565, 1623, 1640, 1641,
    1645, 1655, 1693, 1702, 1867, 1989, 2067, 2083, 2135, 2327, 2357, 2361, 2428,
    2481, 2501, 2555, 2574, 2575, 2585, 2611, 2657, 2677, 2737, 2761, 2787, 2795,
    2817, 2836, 2862, 2877, 2884, 2939, 2970, 2986, 3041, 3047, 3048, 3053, 3100,
    3106, 3131, 3139, 3148, 3160, 3167, 3293, 3393, 3487, 3505, 3537, 3557, 3633,
    3635, 3744, 3783, 3796, 3887, 3903, 3952, 3974, 4003, 4041, 4109, 4137, 4188,
    4380, 4387, 4420, 4457, 4515, 4518, 4673, 4684, 4723, 4729, 4822, 4832, 4863,
    4864, 4893, 4961, 4978, 5034, 5139, 5186, 5211, 5221, 5278, 5289, 5435, 5441,
    5517, 5547, 5682, 5728, 5756, 5807, 5962, 5967, 6009, 6010, 6137, 6321, 6567

Stomatal aperture, diurnal course 84, 284, 459, 524, 956, 984, 1058, 1090, 1159,
    1289, 1319, 1320, 1339, 1351, 1471, 1508, 1513, 1560, 1631, 1798, 1800, 1802,
    2105, 2172, 2213, 2217, 2302, 2677, 2691, 2737, 2787, 2795, 2836, 2970, 3024,
    3041, 3067, 3144, 3351, 3392, 3497, 3634, 3635, 3893, 3970, 4137, 4194, 4204,
    4266, 4268, 4313, 4315, 4316, 4317, 4360, 4380, 4420, 4588, 4897, 4902, 4997,
    5137, 5145, 5547, 5606, 5753, 5855, 5858, 5879, 5958, 5959, 5962, 5994, 6033,
    6118, 6226, 6232, 6321, 6428, 6429

Stomatal aperture, methods, diffusion porometers 84, 284, 308, 397, 474, 612, 837,
    956, 1057, 1058, 1180, 1195, 1336, 1345, 1402, 1500, 1513, 1559, 1621, 1641,
    1643, 1702, 1713, 2282, 2557, 2700, 2701, 2778, 2981, 3006, 3013, 3047, 3107,
    3126, 3887, 4141, 4464, 4608, 5027, 5138, 5210, 5451, 5780, 5907, 6025

Stomatal aperture, methods, direct observations 196, 1564, 1713, 2327, 2444, 3013,
    3070, 4943, 5365, 5660

Stomatal aperture, methods, epidermal strips 189, 361, 393, 648, 773, 774, 775, 786,
    852, 1289, 1513, 1564, 1713, 1792, 1830, 2526, 3321, 3439, 3622, 5190, 5957,
    6204, 6217

Stomatal aperture, methods, infiltration 984, 1180, 1513, 1643, 1713, 3013, 3047,
    4204, 5753, 5754

Stomatal aperture, methods, microrelief methods 57, 788, 1089, 1513, 1564, 1591, 1713,
    3413, 4910, 5365, 6243

Stomatal aperture, methods, other than above and below 255, 293, 679, 1280, 1713,
    2485, 2575, 3546, 4415, 5451, 6337

Stomatal aperture, methods, viscous flow rate porometers 1180, 1356, 1513, 1559, 1565,
    1645, 1713, 1835, 2208, 2777, 2778, 3013, 3067, 4464, 5594, 6025, 6338

Stomatal aperture, oscillations 97, 98, 788, 810, 1319, 1320, 1513, 1800, 2217, 2801,
    2862, 3061, 4003, 4931, 5548, 5994, 6321

Stomatal aperture, seasonal course 1356, 5550, 5959, 6428

Stomatal conductance see Conductance for water vapour and CO_2 transfer, stomata

Stomatal reactivity during leaf ontogenesis 126, 765, 1560, 1645, 1724, 3585, 5507,
    6286

Stomatal types 47, 48, 361, 398, 399, 602, 628, 632, 644, 645, 703, 709, 775, 952,
    1042, 1091, 1476, 1513, 1576, 1592, 1625, 1705, 1913, 1972, 1973, 1974, 2138,
    2226, 2281, 2712, 2783, 3784, 2785, 2868, 2984, 3164, 3190, 3322, 3325, 3384,
    3422, 3621, 3649, 3732, 3741, 3748, 3818, 3892, 4005, 4077, 4105, 4106, 4134,
    4135, 4173, 4255, 4304, 4368, 4379, 4440, 4503, 4639, 4641, 4814, 5029, 5030,
    5040, 5098, 5257, 5258, 5323, 5368, 5477, 5572, 5636, 5709, 5803, 5804, 5890,
    6073, 6149, 6157, 6233, 6321

Structure of cuticle 8, 183, 344, 573, 1045, 1117, 1231, 1240, 1408, 1483, 1751,
    1755, 1901, 2081, 2082, 2166, 2167, 2350, 2375, 2574, 2783, 2785, 2982, 2999,

Transpiration cuticular 576, 1045, 1338, 1342, 1378, 1830, 1908, 2166, 2350, 2877,
    2931, 3728, 3744, 3789, 3904, 4320, 5924, 6075, 6218, 6311, 6340

Transpiration, effect of antitranspirants see Antitranspirants

Transpiration integrated 17, 23, 46, 104, 193, 198, 203, 223, 227, 233, 242, 245,
    247, 312, 335, 341, 346, 378, 379, 468, 504, 545, 562, 598, 674, 690, 737, 741,
    795, 817, 836, 857, 872, 874, 884, 886, 916, 945, 968, 970, 1036, 1136, 1200,
    1292, 1310, 1321, 1398, 1415, 1530, 1579, 1595, 1621, 1659, 1688, 1692, 1728,
    1741, 1822, 1860, 1876, 1904, 1968, 1984, 2015, 2071, 2113, 2240, 2271, 2295,
    2300, 2454, 2579, 2582, 2618, 2683, 2799, 2895, 2913, 2956, 2975, 2988, 3002,
    3018, 3062, 3072, 3092, 3156, 3157, 3204, 3256, 3287, 3332, 3365, 3537, 3591,
    3713, 3879, 3919, 3989, 4139, 4444, 4497, 4696, 4718, 4738, 4782, 4880, 4997,
    5210, 5454, 5550, 5578, 5659, 6167, 6282, 6385, 6419

Transpiration rate and stomata see Stomata and transpiration rate

Transpiration rate, comparison of plants with different types of carbon metabolisms
    2691, 2789, 3993, 4101, 4390, 5381, 5686

Transpiration rate, diurnal course 18, 40, 41, 98, 190, 203, 258, 278, 279, 312, 318,
    337, 362, 415, 459, 468, 474, 478, 516, 554, 597, 625, 651, 654, 668, 695, 766,
    767, 802, 870, 910, 944, 984, 1070, 1159, 1184, 1196, 1197, 1291, 1300, 1344,
    1347, 1471, 1493, 1578, 1579, 1646, 1736, 1904, 2023, 2030, 2080, 2103, 2215,
    2217, 2233, 2283, 2490, 2545, 2550, 2578, 2636, 2677, 2729, 2802, 2812, 2884,
    2951, 2979, 3005, 3017, 3041, 3048, 3061, 3074, 3093, 3127, 3159, 3205, 3208,
    3257, 3258, 3294, 3342, 3377, 3392, 3393, 3417, 3479, 3505, 3535, 3548, 3569,
    3590, 3620, 3710, 3785, 3787, 3806, 3811, 3832, 3989, 4084, 4150, 4182, 4214,
    4215, 4268, 4314, 4327, 4341, 4416, 4421, 4433, 4446, 4466, 4500, 4527,
    4588, 4595, 4738, 4743, 4763, 4790, 4829, 4859, 4860, 4880, 5004, 5018, 5027,
    5046, 5058, 5081, 5097, 5137, 5186, 5210, 5211, 5342, 5380, 5448, 5454, 5541,
    5546, 5550, 5579, 5625, 5627, 5689, 5769, 5831, 5868, 5870, 5908, 5910, 5922,
    5965, 6025, 6031, 6156, 6167, 6237, 6246, 6419, 6428, 6464, 6506, 6511, 6519,
    6587

Transpiration rate, effect of fine structure 1021

Transpiration rate, effect of leaf temperature 193, 258, 369, 847, 1161, 1409, 1454,
    2425, 2426, 3590, 3765, 6336

Transpiration rate, gradients of air humidity in leaf intercellular spaces 170, 601,
    766, 809, 814, 1648, 1655, 1898, 2776, 3365, 3379, 3380, 3438, 3505, 3765, 3832,
    4981, 5463, 5574, 5909, 6009, 6336

Transpiration rate, heterogeneity of single leaf blade 255, 3600

Transpiration rate in artificial conditions 55, 63, 69, 91, 97, 98, 107, 109, 125,
    151, 159, 161, 172, 183, 184, 190, 197, 201, 203, 209, 219, 225, 233, 234, 246,
    247, 252, 279, 305, 306, 307, 312, 378, 379, 381, 387, 448, 451, 501, 537, 542,
    543, 559, 560, 568, 579, 597, 606, 610, 690, 691, 712, 747, 765, 781, 786, 788,
    809, 848, 857, 878, 911, 918, 921, 933, 970, 987, 1031, 1040, 1042, 1053, 1074,
    1136, 1147, 1159, 1162, 1172, 1181, 1252, 1330, 1337, 1387, 1410, 1424, 1461,
    1477, 1492, 1493, 1503, 1550, 1566, 1567, 1574, 1617, 1653, 1658, 1671, 1683,
    1685, 1686, 1692, 1715, 1727, 1732, 1763, 1769, 1819, 1830, 1871, 1905, 1920,
    1921, 1928, 1936, 1940, 1948, 1970, 1990, 2039, 2063, 2080, 2141, 2187, 2189,
    2215, 2216, 2217, 2229, 2241, 2243, 2263, 2264, 2305, 2307, 2368, 2391, 2418,
    2422, 2440, 2449, 2454, 2464, 2476, 2515, 2518, 2519, 2531, 2545, 2549, 2555,
    2580, 2582, 2589, 2652, 2657, 2664, 2673, 2739, 2766, 2782, 2812, 2858, 2877,
    2923, 2941, 2942, 2952, 2956, 2979, 2986, 3027, 3055, 3072, 3093, 3154, 3157,
    3167, 3177, 3213, 3218, 3219, 3241, 3250, 3258, 3282, 3287, 3293, 3301, 3317,
    3336, 3391, 3417, 3441, 3447, 3475, 3488, 3537, 3547, 3588, 3600, 3629, 3643,
    3656, 3678, 3687, 3689, 3691, 3738, 3739, 3779, 3787, 3791, 3802, 3862, 3865,
    3970, 4035, 4110, 4150, 4182, 4215, 4233, 4234, 4235, 4239, 4273, 4274, 4296,
    4300, 4342, 4345, 4387, 4407, 4416, 4424, 4472, 4505, 4527, 4528, 4594, 4617,

Transpiration rate, theoretical background 27, 28, 50, 258, 333, 341, 362, 370, 380,
    488, 490, 625, 770, 847, 866, 925, 1078, 1102, 1180, 1184, 1214, 1409, 1448,
    1458, 1735, 1994, 2195, 2251, 2255, 2263, 2278, 2279, 2342, 2344, 2364, 2381,
    2394, 2425, 2426, 2446, 2529, 2618, 2648, 2649, 2840, 2945, 2946, 3041, 3062,
    3067, 3086, 3252, 3266, 3468, 3590, 3784, 3806, 3948, 4051, 4267, 4565, 5026,
    5091, 5134, 5297, 5304, 5463, 5577, 5579, 5800, 5868, 5936, 5942, 6031, 6032,
    6047, 6173, 6174, 6228, 6246, 6266, 6334, 6383, 6413, 6415, 6542, 6567

Transpiration stomatal 3728, 3904, 4320, 6281

Transport of water see Water transport ...

Trichomes see Leaf surface, waxes and trichomes

Tritium oxide see D_2O, T_2O, ...

Turgor pressure see Presure potential ...

W

Water absorption and ion uptake 29, 110, 166, 219, 220, 239, 629, 672, 734, 1019,
    1082, 1101, 1228, 1309, 1341, 1375, 1404, 1542, 1562, 1609, 1627, 1653, 1662,
    1699, 1771, 1773, 1778, 1816, 1817, 1874, 1895, 1896, 1897, 2003, 2034, 2086,
    2106, 2130, 2173, 2187, 2189, 2223, 2237, 2290, 2427, 2438, 2473, 2510, 2522,
    2553, 2770, 2780, 2786, 2965, 2971, 3105, 3128, 3186, 3210, 3234, 3284, 3340,
    3458, 3617, 3643, 3967, 3990, 4004, 4091, 4124, 4198, 4376, 4529, 4669, 4760,
    4961, 5146, 5284, 5300, 5445, 5509, 5511, 5669, 5695, 5704, 5773, 5807, 5816,
    5861, 5888, 5906, 5908, 5919, 5934, 5960, 5969, 6032, 6152, 6171, 6239, 6250,
    6260, 6306, 6440, 6481, 6510, 6511, 6512, 6530

Water absorption by discs of leaf tissue 786, 1609, 2132, 5486

Water absorption by parts of plant 572, 639, 744, 867, 1147, 1228, 1623, 2086, 2181,
    2206, 2218, 2374, 3083, 3437, 3617, 3771, 3929, 5249, 5448, 5522, 5687, 5688,
    6121, 6190

Water absorption by plant see Age of plant, ...; Anatomical structure, ...; Anti-
    biotics, ...; Carbohydrates, ...; CO_2, ...; Cultivars, ...; Defoliation, decapi-
    tation, ...; D_2O, T_2O, ...; Drought, ...; Ecotypes, ...; Enzyme inhibitors, ...;
    Enzymes, ...; Farming practices, ...; Flooding, ...; Growth substances, hormones,
    inhibitors, etc, ...; Humidity of air, ...; Irradiance, ...; Irrigation, ...;
    Lipids and fatty acids, ...; Mineral elements, ...; Osmotic agents, ...; Oxygen,
    ...; Pathological effects, ...; Pesticides and herbicides, ...; pH, ...; Pollu-
    tants and ozone, ...; Precipitation and dew, ...; Salinity, ...; Soil moisture,
    ...; Temperature, effect on water absorption by plant

Water absorption by plant, diurnal course 790, 1159, 1268, 1323, 1419, 1866, 1904,
    2093, 2217, 2222, 3145, 3342, 3354, 3529, 3771, 4860, 5448, 5852, 6423, 6481,
    6511

Water absorption by plant, effect of anaerobic conditions 4384, 5511, 5677

Water absorption by plant, oscillations 4618

Water absorption by plant, seasonal course 100, 302, 871, 1382, 1510, 1674, 1723,
    2093, 2445, 2498, 2537, 3752, 4038, 4291, 4530, 4779, 5000, 5072, 5114, 5120,
    5183, 5211, 5284, 5287, 5590, 5906, 5963, 6359, 6515

Water absorption by seed 79, 141, 329, 520, 678, 757, 1312, 1332, 1464, 1520, 1536,
    1628, 1634, 1663, 1667, 1818, 1823, 1858, 1875, 1893, 1937, 1938, 1967, 2000,
    2009, 2010, 2032, 2046, 2047, 2115, 2139, 2170, 2220, 2223, 2225, 2329, 2420,
    2457, 2469, 2480, 2820, 2872, 2874, 2875, 2898, 2980, 3095, 3121, 3172, 3176,
    3240, 3259, 3371, 3398, 3498, 3499, 3502, 3558, 3636, 3663, 3698, 3736, 3762,

Water absorption by seed (continued) 3763, 3798, 3839, 3941, 3944, 4056, 4092,
    4131, 4241, 4568, 4602, 4734, 4752, 4777, 5013, 5060, 5136, 5158, 5159, 5179,
    5304, 5308, 5379, 5443, 5465, 5487, 5618, 5717, 5765, 5766, 5767, 5782, 5834,
    5857, 5983, 5995, 6009, 6055, 6138, 6193, 6276, 6279, 6318, 6474, 6502, 6535,
    6543

Water absorption, effect of anatomical structure 1914, 2107

Water absorption, effect of fine structure 1446, 2804, 2893, 3274

Water absorption from atmosphere 41, 42, 95, 141, 477, 738, 1042, 1301, 1503, 1556,
    1557, 1694, 1759, 1824, 1934, 1982, 2124, 2177, 2260, 2386, 2438, 2804, 2893,
    3274, 3700, 3986, 4296, 4818, 4821, 6461, 6542

Water absorption from soil 62, 99, 103, 146, 168, 182, 220, 287, 302, 312, 357, 410,
    437, 439, 449, 543, 585, 651, 683, 714, 737, 788, 803, 845, 871, 883, 894, 954,
    957, 1012, 1013, 1033, 1070, 1144, 1175, 1207, 1261, 1334, 1360, 1365, 1371,
    1380, 1489, 1533, 1541, 1542, 1569, 1610, 1627, 1653, 1669, 1695, 1736, 1759,
    1781, 1847, 1849, 1897, 1905, 1910, 1970, 2009, 2041, 2093, 2107, 2186, 2217,
    2237, 2240, 2264, 2314, 2337, 2368, 2412, 2424, 2445, 2470, 2473, 2510, 2533,
    2537, 2538, 2572, 2581, 2702, 2733, 2788, 2810, 2893, 2902, 2903, 2906, 2908,
    3031, 3036, 3119, 3123, 3203, 3281, 3342, 3344, 3347, 3354, 3412, 3426, 3459,
    3468, 3542, 3557, 3571, 3616, 3640, 3665, 3778, 3803, 3860, 3864, 3868, 3869,
    3960, 4000, 4038, 4059, 4116, 4124, 4128, 4178, 4183, 4184, 4186, 4192, 4277,
    4291, 4307, 4347, 4384, 4530, 4619, 4660, 4669, 4683, 4737, 4765, 4785, 4839,
    4843, 4865, 4889, 4916, 4933, 4981, 5000, 5069, 5110, 5120, 5165, 5166, 5167,
    5183, 5189, 5252, 5284, 5304, 5320, 5389, 5398, 5461, 5468, 5474, 5494, 5495,
    5509, 5528, 5532, 5577, 5578, 5579, 5628, 5682, 5696, 5704, 5794, 5852, 5906,
    5908, 5945, 5949, 5969, 6011, 6018, 6019, 6025, 6027, 6039, 6148, 6170, 6177,
    6179, 6220, 6221, 6237, 6239, 6249, 6272, 6274, 6331, 6340, 6398, 6418, 6452,
    6456, 6457, 6468, 6481, 6528, 6532, 6534, 6538

Water absorption from solution 106, 107, 121, 141, 219, 734, 790, 843, 986, 1031,
    1085, 1095, 1153, 1154, 1159, 1228, 1285, 1503, 1732, 1759, 1918, 1920, 1970,
    1976, 2040, 2109, 2179, 2187, 2215, 2216, 2217, 2316, 2374, 2396, 2579, 2754,
    2877, 2971, 3055, 3141, 3186, 3282, 3300, 3301, 3302, 3441, 3447, 3617, 3665,
    3771, 3862, 4126, 4345, 4416, 4572, 4573, 4781, 4782, 4783, 4847, 4860, 4976,
    5283, 5284, 5306, 5337, 5889, 5969, 6260, 6511

Water absorption, oscillations 2217, 2222

Water balance of cells and tissues 88, 109, 182, 191, 226, 243, 276, 312, 322, 404,
    570, 583, 666, 780, 785, 894, 1185, 1255, 1259, 1268, 1270, 1400, 1485, 1508,
    1562, 1652, 1777, 1786, 1829, 1859, 2501, 2538, 2679, 2686, 3055, 3139, 3241,
    3294, 3485, 3524, 3712, 4096, 4166, 4192, 4193, 4294, 4366, 4372, 4425, 4478,
    4625, 4861, 4879, 4946, 4984, 5226, 5441, 5606, 5988, 6028, 6303

Water balance of whole plant 33, 58, 128, 148, 193, 297, 543, 666, 679, 714, 813,
    894, 916, 1026, 1027, 1028, 1034, 1060, 1074, 1077, 1080, 1143, 1185, 1240, 1260,
    1264, 1343, 1368, 1371, 1436, 1562, 1645, 1699, 1736, 1740, 1789, 1916, 1920,
    1953, 1969, 1970, 2144, 2145, 2184, 2243, 2275, 2296, 2397, 2398, 2413, 2432,
    2470, 2538, 2614, 2840, 2977, 3011, 3055, 3125, 3203, 3204, 3242, 3342, 3405,
    3440, 3447, 3862, 3872, 3962, 4051, 4166, 4183, 4192, 4353, 4436, 4470, 4528,
    4589, 4745, 4781, 5267, 5283, 5526, 5606, 5869, 5879, 5908, 6032, 6135, 6247,
    6365, 6388, 6389, 6468, 6477, 6537

Water balance of whole plant, methods 1369, 1830, 3342, 5283

Water consumption 17, 49, 65, 85, 103, 142, 169, 206, 235, 263, 302, 349, 357, 493,
    534, 598, 616, 634, 651, 692, 788, 857, 866, 876, 973, 1006, 1069, 1077, 1081,
    1103, 1116, 1140, 1162, 1179, 1197, 1206, 1220, 1263, 1278, 1295, 1335, 1388,
    1392, 1447, 1471, 1502, 1519, 1538, 1580, 1581, 1666, 1682, 1735, 1736, 1753,
    1754, 1790, 1815, 1846, 1904, 1905, 1915, 1963, 1968, 1969, 1970, 1995, 2023,
    2024, 2026, 2037, 2057, 2071, 2093, 2156, 2200, 2211, 2214, 2234, 2256, 2316,

Water consumption (continued) 2380, 2454, 2497, 2498, 2505, 2535, 2563, 2586, 2588,
    2597, 2609, 2616, 2630, 2638, 2650, 2660, 2683, 2685, 2693, 2720, 2774, 2781,
    2796, 2806, 2813, 2856, 2950, 2957, 2965, 2968, 2983, 3009, 3030, 3031, 3032,
    3037, 3052, 3065, 3105, 3123, 3150, 3195, 3205, 3227, 3235, 3243, 3247, 3256,
    3257, 3296, 3305, 3308, 3311, 3335, 3348, 3394, 3420, 3423, 3425, 3455, 3460,
    3461, 3462, 3466, 3484, 3504, 3518, 3532, 3552, 3603, 3614, 3646, 3705, 3707,
    3719, 3721, 3722, 3753, 3764, 3771, 3823, 3826, 3832, 3869, 3886, 3906, 3933,
    3958, 3973, 3985, 4000, 4001, 4020, 4021, 4128, 4140, 4184, 4200, 4217, 4274,
    4337, 4346, 4433, 4460, 4547, 4566, 4629, 4637, 4640, 4648, 4685, 4763, 4775,
    4776, 4782, 4785, 4870, 4887, 4942, 4951, 4970, 4998, 4999, 5000, 5037, 5071,
    5090, 5094, 5183, 5227, 5252, 5290, 5321, 5398, 5444, 5461, 5462, 5463, 5482,
    5501, 5526, 5554, 5628, 5641, 5718, 5719, 5721, 5799, 5831, 5888, 5897, 5918,
    5932, 5934, 5938, 5963, 5986, 6018, 6039, 6052, 6057, 6106, 6134, 6135, 6151,
    6211, 6268, 6271, 6272, 6340, 6381, 6382, 6398, 6456, 6476, 6496, 6515, 6516,
    6526

Water consumption, methods 341, 1293, 1448, 1651, 1806, 1869, 2205, 2745, 2797,
    2914, 3693, 3721, 3722, 3752, 3753, 3760, 3829, 3877, 3900, 3901, 3938, 3960,
    3988, 3990, 4000, 4001, 4002, 4140, 4497, 4507, 4640, 4685, 4697, 4828, 4870,
    4985, 5071, 5134, 5234, 5424, 5580, 5677, 5678, 5697, 6022, 6524

Water content in plant and related volume changes 25, 26, 69, 152, 156, 358, 376,
    570, 639, 728, 737, 740, 867, 883, 899, 962, 1138, 1149, 1185, 1239, 1258, 1331,
    1363, 1400, 1465, 1497, 1638, 1653, 1740, 1805, 1832, 1998, 2029, 2079, 2109,
    2149, 2172, 2205, 2441, 2442, 2476, 2745, 2746, 2826, 2929, 2968, 3061, 3079,
    3082, 2083, 3110, 3260, 3294, 3354, 3376, 3415, 3587, 3685, 3860, 3906, 3981,
    4073, 4088, 4267, 4366, 4452, 4670, 4742, 4746, 4875, 4928, 5055, 5164, 5211,
    5448, 5662, 5787, 5879, 5989, 6143, 6388, 6389

Water content in plant, methods 25, 164, 228, 272, 273, 296, 494, 639, 671, 737, 853,
    901, 939, 964, 1106, 1198, 1218, 1638, 2029, 2205, 2207, 2323, 2543, 2929, 3013,
    3045, 3491, 3560, 3694, 4049, 4136, 4233, 4234, 4235, 4939, 5785, 5802, 5995,
    6025, 6488

Water content in plant tissues 5, 24, 32, 38, 73, 83, 87, 95, 103, 104, 106, 113,
    135, 158, 164, 165, 167, 183, 208, 210, 213, 249, 259, 262, 268, 269, 272, 275,
    276, 278, 297, 312, 355, 356, 396, 400, 401, 403, 404, 419, 422, 431, 432, 446,
    455, 459, 460, 462, 463, 466, 468, 477, 484, 491, 520, 533, 535, 536, 547, 552,
    577, 583, 588, 610, 664, 671, 674, 677, 678, 682, 695, 697, 704, 705, 706, 737,
    738, 741, 744, 749, 756, 780, 794, 796, 799, 803, 824, 838, 848, 874, 879, 891,
    892, 894, 905, 907, 929, 940, 957, 971, 973, 1005, 1014, 1021, 1030, 1031, 1050,
    1071, 1100, 1120, 1137, 1146, 1173, 1213, 1218, 1221, 1223, 1236, 1239, 1250,
    1290, 1291, 1328, 1331, 1341, 1344, 1352, 1357, 1358, 1368, 1386, 1400, 1411,
    1437, 1439, 1443, 1452, 1463, 1479, 1480, 1485, 1493, 1525, 1537, 1539, 1540,
    1571, 1583, 1611, 1631, 1638, 1645, 1646, 1657, 1673, 1677, 1703, 1738, 1765,
    1785, 1795, 1796, 1823, 1824, 1837, 1855, 1859, 1862, 1870, 1871, 1872, 1877,
    1883, 1892, 1893, 1906, 1927, 1928, 1929, 1951, 1954, 1968, 1991, 1993, 1997,
    1998, 2004, 2025, 2033, 2077, 2101, 2114, 2115, 2120, 2123, 2125, 2133, 2142,
    2144, 2145, 2147, 2152, 2155, 2169, 2200, 2207, 2224, 2229, 2236, 2250, 2266,
    2287, 2289, 2293, 2294, 2295, 2312, 2314, 2321, 2323, 2332, 2333, 2350, 2352,
    2372, 2396, 2416, 2430, 2449, 2456, 2462, 2471, 2508, 2509, 2514, 2539, 2550,
    2555, 2564, 2567, 2596, 2614, 2636, 2643, 2648, 2649, 2652, 2676, 2686, 2692,
    2715, 2716, 2726, 2746, 2747, 2820, 2824, 2827, 2852, 2863, 2874, 2875, 2881,
    2912, 2919, 2920, 2943, 2953, 2990, 2992, 2993, 3003, 3016, 3018, 3028, 3071,
    3112, 3128, 3169, 3188, 3201, 3231, 3240, 3241, 3257, 3279, 3284, 3362, 3371,
    3403, 3416, 3448, 3468, 3477, 3485, 3510, 3526, 3532, 3550, 3561, 3571, 3580,
    3581, 3582, 3583, 3584, 3628, 3629, 3630, 3644, 3646, 3653, 3654, 3675, 3688,
    3694, 3711, 3724, 3737, 3740, 3743, 3763, 3772, 3782, 3815, 3820, 3825, 3833,
    3857, 3874, 3876, 3890, 3894, 3907, 3911, 3915, 3941, 3963, 3989, 4009, 4010,
    4032, 4036, 4042, 4047, 4049, 4070, 4071, 4073, 4074, 4084, 4126, 4136, 4145,
    4177, 4179, 4222, 4227, 4228, 4233, 4234, 4235, 4236, 4240, 4241, 4243, 4249,
    4262, 4268, 4270, 4276, 4278, 4283, 4291, 4293, 4295, 4296, 4310, 4316, 4327,
    4328, 4336, 4347, 4362, 4367, 4387, 4388, 4393, 4394, 4395, 4397, 4400, 4402,
    4405, 4411, 4425, 4426, 4432, 4433, 4441, 4448, 4449. 4450, 4451, 4453, 4473,

Water potential, methods, pressure bomb   10, 129, 130, 474, 542, 555, 556, 633, 728,
      763, 877, 913, 1067, 1140, 1219, 1316, 1399, 1401, 1529, 1665, 1713, 2031, 2095,
      2360, 2363, 2421, 2777, 2778, 3013, 3029, 3144, 3355, 3356, 3417, 3419, 3653,
      4094, 4301, 4314, 4808, 5107, 5162, 5164, 5786, 5809, 5908, 6025, 6240, 6254,
      6338, 6389, 6563

Water potential, methods, psychrometry   150, 187, 548, 556, 614, 633, 668, 823, 859,
      877, 1068, 1156, 1316, 1399, 1550, 1713, 1824, 2069, 2095, 2421, 2437, 2511,
      2954, 3013, 3029, 4314, 4646, 4647, 4827, 5104, 5908, 6025

Water potential, methods, psychrometry in situ 1157, 1399, 1492, 1881, 2595, 3103,
      3226, 3602, 3669, 6021, 6387, 6552

Water saturation deficit 2, 4, 40, 58, 63, 77, 86, 128, 133, 139, 144, 186, 188, 198,
      199, 203, 213, 223, 225, 263, 268, 278, 279, 312, 335, 336, 355, 365, 376, 378,
      404, 463, 465, 466, 468, 470, 471, 472, 474, 475, 478, 483, 484, 487, 507, 537,
      538, 556, 557, 563, 572, 574, 580, 606, 611, 615, 617, 631, 673, 679, 687, 688,
      728, 741, 750, 755, 766, 769, 784, 787, 810, 812, 845, 894, 899, 929, 955, 972,
      973, 992, 995, 1013, 1023, 1031, 1088, 1106, 1125, 1180, 1290, 1301, 1314, 1316,
      1318, 1321, 1330, 1337, 1339, 1345, 1347, 1348, 1349, 1367, 1371, 1373, 1395,
      1437, 1438, 1447, 1503, 1507, 1532, 1547, 1562, 1577, 1597, 1620, 1632, 1650,
      1676, 1683, 1687, 1710, 1728, 1736, 1776, 1779, 1782, 1819, 1822, 1834, 1851,
      1871, 1920, 1927, 1929, 1934, 1978, 1991, 2019, 2028, 2066, 2080, 2092, 2148,
      2200, 2203, 2204, 2211, 2217, 2242, 2250, 2293, 2305, 2362, 2363, 2386, 2397,
      2398, 2412, 2440, 2484, 2487, 2501, 2519, 2520, 2528, 2545, 2546, 2549, 2550,
      2601, 2677, 2680, 2682, 2684, 2697, 2729, 2735, 2741, 2757, 2808, 2819, 2831,
      2844, 2863, 2865, 2933, 2937, 2947, 2951, 2987, 2989, 2993, 3025, 3028, 3029,
      3055, 3067, 3113, 3148, 3166, 3213, 3292, 3312, 3320, 3348, 3356, 3388, 3390,
      3405, 3477, 3488, 3532, 3544, 3548, 3565, 3632, 3653, 3662, 3666, 3706, 3710,
      3734, 3779, 3782, 3814, 3815, 3819, 3832, 3864, 3868, 3894, 3962, 3963, 3977,
      3991, 4027, 4041, 4042, 4057, 4064, 4066, 4104, 4118, 4152, 4210, 4224, 4225,
      4226, 4227, 4234, 4268, 4316, 4321, 4326, 4327, 4328, 4335, 4342, 4347, 4353,
      4356, 4371, 4397, 4478, 4548, 4562, 4569, 4638, 4671, 4672, 4674, 4677, 4712,
      4762, 4767, 4798, 4810, 4812, 4818, 4832, 4835, 4849, 4853, 4854, 4861, 4880,
      4893, 4896, 4911, 4928, 4939, 4942, 4943, 4947, 4948, 4965, 4975, 4982, 5009,
      5048, 5103, 5105, 5106, 5127, 5164, 5195, 5204, 5283, 5304, 5305, 5320, 5400,
      5463, 5546, 5550, 5551, 5607, 5611, 5639, 5654, 5688, 5694, 5730, 5746, 5762,
      5769, 5786, 5787, 5815, 5822, 5849, 5858, 5905, 5965, 5969, 6025, 6027, 6032,
      6069, 6088, 6156, 6212, 6213, 6215, 6222, 6307, 6313, 6336, 6339, 6391, 6393,
      6415, 6427, 6428, 6436, 6471, 6495, 6532, 6542, 6563, 6568, 6569, 6583, 6584,
      6592

Water saturation deficit, methods   150, 355, 365, 494, 571, 572, 611, 1713, 2472,
      2741, 3001, 3013, 3029, 3146, 3546, 5283, 5806, 6338

Water status in plant see also Age of plant, ...; Altitude and pressure, ...; Carbo-
      hydrates, ...; Cultivars, ...; Defoliation, decapitation, ...; Drought, ...;
      Ecotypes, ...; Enzyme inhibitors, ...; Enzymes, ...; Farming practices, ...;
      Flooding, ...; Genetics, ...; Growth substances, hormones, inhibitors etc., ...;
      Humidity of air, ...; Irradiance, ...; Irrigation, ...; Leaf insertion level,
      ...; Lipids and fatty acids, ...; Mineral elements, ...; Mutants, ...; Osmotic
      agents, ...; Oxygen, ...; Pathological effects, ...; Pesticides and herbicides,
      ...; pH, ...; Pollutants and ozone, ...; Precipitation and dew, ...; Proteins,
      amino acids and nucleic acids, ...; Salinity, ...; Soil moisture, ...; Taxons,
      ...; Temperature, ...; Wind, effect on water status in plant

Water status in plant, comparison of plants with different types of carbon metabolisms
      3994, 4896, 6425

Water status in plant, diurnal course 25, 171, 198, 199, 203, 223, 256, 312, 339,
      340, 352, 376, 400, 401, 404, 411, 432, 436, 459, 462, 468, 472, 474, 478, 522,
      526, 551, 625, 633, 641, 647, 673, 679, 722, 728, 766, 767, 838, 845, 868, 908,
      910, 971, 972, 1028, 1043, 1109, 1138, 1185, 1196, 1235, 1300, 1339, 1347, 1363,
      1400, 1502, 1529, 1562, 1563, 1613, 1633, 1646, 1702, 1802, 1805, 1808, 1845,

Water status in plant, diurnal course (continued) 1851, 1940, 2030, 2072, 2079,
    2080, 2095, 2105, 2109, 2117, 2148, 2200, 2201, 2205, 2215, 2245, 2250, 2269,
    2276, 2330, 2431, 2432, 2484, 2511, 2545, 2562, 2583, 2595, 2601, 2619, 2655,
    2665, 2677, 2682, 2697, 2754, 2844, 2888, 2900, 2902, 2929, 2933, 2934, 2935,
    2955, 3041, 3061, 3064, 3079, 3107, 3110, 3127, 3131, 3133, 3144, 3159, 3180,
    3222, 3257, 3260, 3261, 3263, 3287, 3320, 3350, 3354, 3393, 3403, 3405, 3417,
    3433, 3479, 3486, 3487, 3563, 3634, 3635, 3642, 3658, 3677, 3710, 3742, 3744,
    3806, 3860, 3871, 3889, 3906, 3921, 3966, 3975, 4025, 4026, 4045, 4062, 4064,
    4065, 4150, 4182, 4194, 4266, 4267, 4268, 4296, 4297, 4306, 4307, 4316, 4327,
    4353, 4380, 4453, 4460, 4500, 4530, 4588, 4659, 4660, 4662, 4690, 4742, 4767,
    4768, 4829, 4860, 4861, 4880, 4902, 5003, 5010, 5018, 5025, 5046, 5054, 5069,
    5073, 5105, 5204, 5245, 5318, 5448, 5486, 5541, 5550, 5551, 5556, 5558, 5579,
    5588, 5606, 5624, 5625, 5627, 5649, 5708, 5727, 5744, 5745, 5769, 5788, 5792,
    5806, 5844, 5858, 5870, 5965, 6032, 6122, 6132, 6143, 6156, 6237, 6255, 6257.
    6291, 6297, 6312, 6317, 6333, 6374, 6376, 6387, 6391, 6428, 6533, 6537, 6559,
    6587.

Water status in plant, effect on biliproteins 3028

Water status in plant, effect on canopy architecture 1133, 1314, 6326

Water status in plant, effect on carbon fixation pathways 509, 869, 903, 1133, 1568,
    1689, 1733, 1745, 1946, 1948, 2541, 2682, 2987, 3049, 3050, 3219, 3279, 3446,
    3773, 3950, 4143, 4213, 4280, 4359, 4417, 4420, 4482, 4571, 4607, 4920, 5156,
    5350, 5543, 5908, 6231, 6484

Water status in plant, effect on carotenoids 1222, 1325, 1326, 1757, 4450, 5009,
    6402

Water status in plant, effect on chlorophyll 15, 96, 407, 465, 501, 588, 769, 960,
    993, 1134, 1222, 1325, 1326, 1525, 1562, 1857, 2027, 2028, 2210, 2384, 2494,
    2541, 2846, 2987, 3028, 3201, 3293, 3788, 3801, 3832, 3931, 3983, 4046, 4143,
    4280, 4441, 4450, 4574, 5002, 5009, 5063, 5193, 5305, 5524, 5543, 5614, 5677,
    5908, 6012, 6032, 6402

Water status in plant, effect on chloroplasts 229, 310, 460, 462, 465, 664, 740, 745,
    853, 926, 960, 1134, 1326, 1396, 1418, 1423, 2007, 2384,2947, 3152, 3207, 3293,
    3800, 3913, 4046, 4088, 4144, 4201, 4321, 4431, 4450, 4517, 4661, 4874, 4877,
    5305, 5447, 5687, 5688, 5758, 5908, 6433, 6589,

Water status in plant, effect on CO_2 influx 46, 63, 91, 95, 96, 104, 171, 209, 221,
    382, 396, 400, 402, 412, 446, 450, 460, 477, 478, 482, 483, 484, 486, 487, 492,
    506, 514, 537, 538, 544, 563, 566, 574, 613, 615, 661, 666, 673, 714, 720, 769,
    781, 787, 802, 810, 861, 902, 904, 905, 993, 1035, 1043, 1074, 1088, 1133, 1134,
    1148, 1169, 1200, 1208, 1240, 1290, 1318, 1338, 1414, 1503, 1505, 1562, 1563,
    1566, 1567, 1595, 1597, 1598, 1622, 1633, 1646, 1676, 1689, 1733, 1746, 1775,
    1776, 1819, 1851, 1857, 1927, 1940, 1948, 1970, 2027, 2148, 2175, 2181, 2211,
    2276, 2299, 2414, 2432, 2482, 2531, 2541, 2583, 2657, 2663, 2668, 2698, 2841,
    2862, 2888, 2905, 2924, 2987, 2993, 3011, 3041, 3042, 3049, 3068, 3099, 3127,
    3148, 3160, 3173, 3219, 3254, 3287, 3293, 3349, 3475, 3488, 3543, 3678, 3773,
    3774, 3779, 3799, 3805, 3837, 3856, 3874, 3882, 3888, 3950, 3959, 3983, 4019,
    4086, 4144, 4201, 4321, 4347, 4354, 4380, 4390, 4392, 4393, 4394, 4395, 4458,
    4470, 4475, 4481, 4482, 4570, 4574, 4589, 4598, 4600, 4774, 4829, 4883, 4920,
    4985, 5054, 5056, 5063, 5083, 5103, 5105, 5156, 5371, 5373, 5381, 5383, 5388,
    5395, 5458, 5467, 5469, 5479, 5543, 5545, 5603, 5659, 5684, 5687, 5698, 5758,
    5788, 5806, 5826, 5900, 5905, 5916, 5920, 5925, 5926, 5943, 5944, 5953, 6000,
    6032, 6059, 6089, 6118, 6261, 6277, 6312, 6313, 6425, 6439, 6441, 6465, 6471,
    6473

Water status in plant, effect on conductance for water vapour and CO_2 transfer 58,
    69, 171, 186, 223, 309, 376, 391, 415, 450, 451, 486, 487, 537, 538, 551, 556,
    574, 606, 615, 673, 720, 730, 766, 767, 781, 787, 810, 908, 993, 1043, 1058,
    1109, 1155, 1169, 1180, 1235, 1300, 1321, 1337, 1363, 1409, 1493, 1531, 1563,
    1620, 1654, 1701, 1732, 1733, 1747, 1775, 1776, 1832, 1845, 1948, 1990, 2027,

Water status in plant, effect on conductance for water vapour and CO_2 transfer (continued) 2128, 2148, 2204, 2208, 2243, 2276, 2328, 2339, 2405, 2412, 2422, 2482, 2483, 2485, 2531, 2541, 2603, 2605, 2652, 2680, 2697, 2735, 2737, 2761, 2844, 2862, 2892, 2923, 2937, 2956, 3041, 3107, 3110, 3127, 3160, 3167, 3219, 3293, 3405, 3442, 3475, 3488, 3489, 3563, 3564, 3635, 3677, 3689, 3773, 3774, 3779, 3868, 3882, 3893, 3933, 3975, 3976, 4089, 4101, 4144, 4266, 4267, 4274, 4306, 4347, 4380, 4390, 4482, 4589, 4600, 4601, 4671, 4690, 4742, 4767, 4772, 4773, 4774, 4829, 4872, 5010, 5054, 5056, 5057, 5103, 5104, 5186, 5210, 5211, 5311, 5341, 5383, 5406, 5413, 5435, 5463, 5467, 5542, 5574, 5579, 5588, 5596, 5603, 5744, 5783, 5787, 5788, 5807, 5849, 5870, 5905, 5923, 5958, 5965, 6000, 6015, 6026, 6027, 6032, 6041, 6059, 6130, 6209, 6216, 6230, 6246, 6255, 6291, 6297, 6311, 6312, 6313, 6321, 6332, 6340, 6392, 6441, 6494, 6583

Water status in plant, effect on electron transport chain 10, 229, 419, 552, 566, 745, 753, 787, 810, 895, 960, 1133, 1364, 1518, 1857, 1919, 2027, 2028, 2055, 2056, 2210, 2494, 2541, 2947, 3160, 3293, 3773, 3801, 3931, 4144, 4425, 4517, 4598, 5193, 5758, 5808, 5891, 6589, 6590

Water status in plant, effect on growth and productivity 1, 91, 96, 141, 165, 223, 317, 352, 402, 484, 512, 533, 571, 616, 652, 678, 722, 799, 821, 836, 993, 1025, 1074, 1121, 1122, 1133, 1134, 1242, 1272, 1358, 1400, 1562, 1563, 1595, 1633, 1775, 1785, 1838, 1918, 1952, 2114, 2269, 2411, 2440, 2465, 2506, 2521, 2532, 2544, 2563, 2604, 2924, 2960, 2987, 2993, 3011, 3041, 3075, 3108, 3110, 3132, 3299, 3368, 3405, 3465, 3502, 3543, 3639, 3753, 3796, 3805, 3815, 3838, 3866, 3889, 3947, 4019, 4047, 4347, 4471, 4570, 4845, 4983, 4985, 5103, 5163, 5356, 5399, 5411, 5469, 5516, 5533, 5585, 5624, 5656, 5684, 5794, 5807, 5837, 5877, 5878, 5879, 5917, 5945, 6006, 6025, 6067, 6168, 6324, 6425, 6440, 6491, 6524, 6596

Water status in plant, effect on leaf anatomy 91, 352, 494, 1134, 1168, 1367, 1562, 1807, 1990, 2541, 2599, 2604, 2655, 2680, 2888, 3133, 3837, 3864, 3866, 3921, 3964, 3971, 4088, 4354, 4686, 5103, 5163, 5321, 5624, 5625, 5684, 5807, 6049, 6168, 6216, 6323, 6332, 6333

Water status in plant, effect on photorespiration 486, 1133, 1169, 1240, 1689, 1746, 2027, 2987, 3049, 3050, 3293, 4390, 4481, 4482, 4748, 4774, 5103, 5603, 5920, 6089

Water status in plant, effect on respiration 295, 400, 412, 419, 423, 477, 482, 486, 714, 720, 781, 826, 848, 926, 993, 1074, 1133, 1134, 1148, 1169, 1562, 1566, 1567, 1676, 1746, 1776, 1857, 1959, 2027, 2268, 2567, 2605, 2888, 3042, 3127, 3169, 3362, 3799, 4390, 4431, 4451, 4569, 4574, 4600, 4774, 5103, 5371, 5373, 5388, 5603, 5684, 5711, 5806, 5891, 5900, 5920, 6032, 6053, 6059, 6089, 6168, 6471

Water status in plant, effect on root anatomy 1428, 2408

Water status in plant, effect on stomata and epidermis 58, 198, 271, 306, 333, 351, 475, 544, 576, 709, 714, 728, 765, 785, 1010, 1058, 1074, 1134, 1255, 1319, 1321, 1329, 1338, 1367, 1396, 1493, 1508, 1513, 1514, 1531, 1560, 1562, 1710, 1800, 1832, 1916, 2060, 2065, 2105, 2217, 2328, 2351, 2485, 2541, 2605, 2652, 2657, 2737, 2761, 2801, 2841, 2862, 2951, 3011, 3040, 3041, 3127, 3144, 3293, 3351, 3441, 3475, 3635, 3689, 3882, 3893, 3952, 3976, 4061, 4089, 4103, 4267, 4347, 4380, 4457, 4515, 4616, 4773, 4943, 5104, 5105, 5186, 5383, 5413, 5415, 5542, 5596, 5615, 5682, 5709, 5788, 5879, 5908, 5957, 5958, 5962, 5975, 6009, 6041, 6118, 6321, 6333, 6392, 6494, 6519, 6566, 6583

Water status in plant, effect on transpiration 63, 160, 161, 183, 184, 217, 308, 318, 376, 401, 450, 544, 606, 695, 766, 767, 802, 810, 822, 848, 874, 915, 1030, 1088, 1196, 1337, 1338, 1508, 1566, 1567, 1646, 1653, 1655, 1738, 1819, 1934, 1936, 1948, 1990, 2030, 2217, 2243, 2276, 2325, 2418, 2432, 2440, 2481, 2485, 2531, 2545, 2549, 2648, 2649, 2652, 2657, 2702, 2923, 2979, 3011, 3018, 3067, 3068, 3109, 3148, 3179, 3219, 3293, 3294, 3488, 3505, 3689, 3779, 3789, 3856, 3864, 3865, 4084, 4201, 4239, 4300, 4353, 4392, 4544, 4829, 4951, 5018, 5056, 5148,

Water status in plant, effect on transpiration (continued) 5195, 5210, 5228, 5336,
    5373, 5413, 5415, 5463, 5467, 5545, 5728, 5781, 5915, 6059, 6088, 6230, 6297,
    6312, 6313, 6340, 6461

Water status in plant, effect on water absorption by plant 1301

Water status in plant, effect on water transport in cells 457, 2166, 3178, 3233,
    4088, 5020, 6311

Water status in plant, effect on water transport in plant 810, 1043, 1562, 1747,
    1940, 2117, 2119, 2612, 4182, 4465, 5969

Water status in plant, heterogeneity of single leaf blade 542, 1213, 1782, 2149,
    2204, 2431, 2885, 3104, 3642, 3863, 4074, 4218, 4410, 4561, 4875, 4876, 4880

Water status in plant, oscillations 538, 1550, 2217, 2431, 3061, 4670,

Water status in plant, seasonal course 94, 144, 156, 165, 171, 198, 199, 327, 355,
    356, 401, 428, 436, 455, 462, 521, 633, 674, 679, 685, 741, 812, 845, 865, 883,
    915, 972, 1048, 1050, 1138, 1140, 1142, 1173, 1203, 1206, 1235, 1274, 1318, 1330,
    1347, 1349, 1352, 1370, 1401, 1443, 1529, 1562, 1577, 1602, 1650, 1728, 1803,
    1838, 1862, 1935, 1951, 1968, 2072, 2184, 2250, 2294, 2298, 2350, 2462, 2484,
    2507, 2601, 2619, 2623, 2626, 2665, 2715, 2735, 2795, 2808, 2900, 2933, 2955,
    3012, 3064, 3108, 3113, 3134, 3159, 3160, 3260, 3350, 3356, 3362, 3390, 3394,
    3486, 3497, 3544, 3549, 3593, 3658, 3675, 3676, 3677, 3710, 3711, 3742, 3761,
    3772, 3774, 3796, 3860, 3871, 3927, 4016, 4042, 4089, 4267, 4283, 4291, 4307,
    4321, 4326, 4327, 4433, 4548, 4575, 4579, 4589, 4592, 4657, 4659, 4660, 4716,
    4732, 4738, 4767, 4768, 4808, 4830, 4855, 4974, 5034, 5057, 5105, 5106, 5188,
    5195, 5318, 5386, 5387, 5428, 5475, 5491, 5493, 5495, 5508, 5558, 5564, 5588,
    5611, 5627, 5642, 5649, 5724, 5745, 5806, 5892, 6016, 6026, 6027, 6049, 6087,
    6088, 6108, 6109, 6132, 6156, 6255, 6275, 6317, 6428, 6477, 6545, 6588

Water status in plant, theoretical background, terminology 45, 296, 497, 663, 900,
    1970, 1996, 3345, 3407, 4203, 4267, 5089, 5164, 5267, 5745, 6025

Water stress development see Altitude and pressure, ...; Drought, effect on water
    stress development

Water stress in plant see Wilting, ...

Water transport in cells (see also Age of plant, ...; Carbohydrates, ...; CO_2, ...;
    Cultivars, ...; D_2O, T_2O, ...; Drought, ...; Enzyme inhibitors, ...; Growth
    substabces, hormones, inhibitors etc., ...; Irradiance, ...; Mineral elements,
    ...; Osmotic agents, ...; Pesticides and herbicides, ...; pH, ...; Pollutants
    and ozone, ...; Salinity, ...; Soil moisture, ...; Temperature, ...; Water status
    in plant, effect on water transport in cells) 32, 331, 433, 434, 549, 609, 695,
    713, 833, 845, 867, 899, 974, 976, 1246, 1270, 1496, 1527, 1600, 1752, 1777,
    1967, 2011, 2018, 2068, 2154, 2166, 2168, 2191, 2284, 2359, 2400, 2449, 2451,
    2478, 2513, 2764, 2765, 2770, 2901, 2917, 2997, 3283, 3415, 3500, 3524, 3626,
    3673, 3680, 3726, 3727, 3751, 3929, 4034, 4129, 4237, 4294, 4373, 4413, 4605,
    4794, 5241, 5242, 5243, 5357, 5455, 5470, 5647, 5655, 5701, 5816, 5850, 5969,
    6025, 6310, 6462, 6474, 6475, 6498, 6522, 6585

Water transport in cells, cell wall structure, modulus of elasticity 129, 130, 147,
    152, 153, 236, 301, 570, 975, 976, 1151, 1219, 1231, 1246, 1270, 1296, 1449,
    1620, 1742, 1777, 1967, 2377, 2671, 2755, 2772, 2901, 3178, 3334, 3509, 3626,
    3680, 3747, 3929, 4294, 4356, 4371, 4485, 4557, 4562, 4708, 4739, 4872, 4882,
    4962, 5106, 5107, 5164, 5241, 5243, 5395, 5400, 5662, 5712, 5969, 6025, 6043,
    6498, 6602

Water transport in cells, membrane structure 469, 627, 1038, 1186, 1246, 1247, 1374,
    1514, 1752, 1777, 1784, 1879, 1967, 2017, 2043, 2130, 2191, 2284, 2451, 2671,
    2755, 2786, 2832, 2901, 2917, 3175, 3334, 3672, 3916, 3929, 4088, 4113, 4290,
    4413, 4605, 4705, 4708, 4879, 4962, 5017, 5324, 5919, 5969, 6298, 6591

Water transport in cells, methods   975, 1752, 2068, 2168, 2401, 3232, 3233, 3524,
    3618, 3725, 4149, 4294, 5035, 5243, 6025, 6028, 6602

Water transport in cells, permeability   32, 153, 296, 325, 433, 434, 456, 457, 520,
    570, 571, 572, 600, 626, 627, 681, 700, 833, 834, 867, 974, 975, 976, 1038, 1246,
    1247, 1270, 1296, 1752, 1777, 1910, 2017, 2068, 2126, 2166, 2167, 2168, 2191,
    2284, 2359, 2400, 2451, 2478, 2496, 2765, 2772, 2917, 3215, 3228, 3229, 3230,
    3232, 3233, 3283, 3458, 3500, 3509, 3524, 3618, 3626, 3680, 3681, 3726, 3838,
    3911, 3912, 3916, 4096, 4149, 4260, 4294, 4413, 4705, 4708, 4746, 4769, 4826,
    4878, 4879, 4949, 5020, 5035, 5036, 5241, 5243, 5304, 5324, 5655, 5662, 5681,
    5885, 5969, 6025, 6028, 6043, 6060, 6250, 6298, 6310, 6311, 6336, 6340, 6417,
    6475, 6522, 6543, 6596, 6602

Water transport in cells, plasmolysis   109, 228, 600, 834, 906, 976, 1420, 1777, 1811,
    2197, 2259, 2363, 2513, 2552, 3297, 3315, 3356, 3661, 3916, 3945, 4034, 4081,
    4096, 4142, 4149, 4195, 4558, 4590, 4616, 4704, 4705, 4707, 4794, 4878, 4956,
    4962, 5106, 5175, 5185, 5220, 5639, 5675, 5820, 5969, 6025, 6028, 6585

Water transport in cells, seasonal course   681

Water transport in plant see also   Age of plant, ...; Anatomical structure, ...;
    CO_2, ...; Cultivars, ...; D_2O, T_2O, ...; Drought, ...; Ecotypes, ...; Flooding,
    ...; Genetics, ...; Growth substances, hormones, inhibitors etc., ...; Humidity
    of air, ...; Irradiance, ...; Irrigation, ...; Leaf insertion level, ...; Mineral
    elements, ...; Osmotic agents, ...; Oxygen, ...; Pathological effects, ...; Pes-
    ticides and herbicides, ...; pH, ...; Pollutants and ozone, ...; Precipitation
    and dew, ...; Proteins, amino acids and nucleic acids, ...; Salinity, ...; Soil
    moisture, ...; Taxons, ...; Temperature, ...; Water status in plant, effect on
    water transport in plant

Water transport in plant, capacitances   110, 376, 737, 739, 842, 921, 1613, 1732,
    1864, 2021, 2109, 2206, 2215, 2217, 2410, 2412, 2654, 2771, 3141, 3865, 3962,
    4267, 4353, 5304, 5569, 5662, 5969, 6032, 6043, 6254

Water transport in plant, conductances   61, 62, 109, 154, 184, 237, 239, 250, 312,
    376, 401, 406, 437, 475, 487, 572, 577, 729, 786, 810, 834, 841, 883, 896, 913,
    921, 942, 1013, 1049, 1096, 1135, 1190, 1191, 1196, 1316, 1375, 1376, 1441, 1450,
    1485, 1528, 1563, 1613, 1695, 1732, 1736, 1747, 1750, 1834, 1840, 1841, 1864,
    1874, 1907, 1910, 1911, 1920, 1930, 2080, 2095, 2106, 2117, 2187, 2206, 2215,
    2217, 2258, 2264, 2363, 2396, 2410, 2418, 2432, 2540, 2581, 2612, 2637, 2757,
    2771, 2780, 2852, 2979, 3141, 3354, 3440, 3473, 3524, 3525, 3588, 3771, 3827,
    3865, 3875, 3962, 4267, 4300, 4331, 4353, 4465, 4582, 4601, 4660, 4722, 4760,
    4837, 4861, 4895, 4901, 4916, 4994, 5237, 5283, 5304, 5389, 5486, 5579, 5595,
    5662, 5771, 5908, 5969, 5992, 6009, 6019, 6025, 6027, 6032, 6223, 6254, 6537

Water transport in plant, diurnal course   577, 651, 1211, 2205, 3022, 3257, 3751,
    4582, 4660, 4940, 6314

Water transport in plant, evaporation sites and comparison of liquid and gaseous
    phases   369, 738, 1032, 1189, 1759, 1831, 1930, 2206, 2262, 2418, 6336

Water transport in plant, heterogeneity of single leaf blade   2218

Water transport in plant, methods   37, 184, 301, 437, 841, 1032, 1171, 1375, 1840,
    2012, 2119, 3141, 3524, 4096, 4901, 5521

Water transport in plant, oscillations   843, 1151

Water transport in plant, radial transport in tree stems   376, 639, 729, 833, 862,
    1334, 1750, 1773, 2130, 2290, 2412, 3475, 3525, 5885, 5969, 6388

Water transport in plant, seasonal course   1278, 2630, 4837

Water transport in plant, theoretical background 29, 35, 38, 45, 109, 154, 158, 237,
    239, 295, 376, 475, 613, 626, 633, 683, 841, 913, 942, 974, 1139, 1191, 1311,
    1432, 1450, 1653, 1742, 1759, 1778, 1784, 1840, 1841, 1864, 1902, 1930, 2119,
    2214, 2262, 2363, 2396, 2400, 2401, 2410, 2772, 3010, 3209, 3751, 3967. 4267,
    4590, 4619, 4707, 5093, 5236, 5237, 5243, 5405, 5969, 6009, 6512

Water transport in plant, transport in leaf 37, 230, 309, 325, 376, 475, 539, 571,
    731, 786, 913, 1032, 1151, 1171, 1189, 1299, 1439, 1473, 1618, 1742, 1778, 1830,
    1832, 2012, 2117, 2135, 2206, 2218, 2262, 2418, 2525, 2581, 2654, 2679, 2705,
    2731, 2902, 3010, 3139, 3162, 3282, 3475, 3524, 3588, 3884, 4096, 4299, 5226,
    5336, 5337, 5405, 5486, 5953, 5969, 5976, 5977, 5978, 5988, 6025, 6043, 6254,
    6321, 6336, 6340, 6343, 6494

Water transport in plant, transport in other organs than above and below 38, 231,
    867, 1049, 1439, 1473, 1778, 2654, 2705, 2771, 3944, 4341, 5662

Water transport in plant, transport in roots 29, 37, 106, 184, 238, 239, 301, 376,
    406, 475, 548, 577, 633, 764, 790, 841, 842, 845, 851, 883, 942, 1019, 1070,
    1191, 1299, 1334, 1375, 1407, 1450, 1473, 1553, 1563, 1613, 1620, 1732, 1742,
    1773, 1777, 1778, 1840, 1841, 1850, 1859, 1874, 1910, 1911, 1920, 2080, 2106,
    2107, 2117, 2187, 2201, 2258, 2262, 2290, 2401, 2418, 2470, 2540, 2612, 2705,
    2728, 2779, 2780, 2810, 2834, 3010, 3141, 3282, 3353, 3447, 3475, 3557, 3667,
    3875, 3910, 3933, 4096, 4183, 4267, 4287, 4372, 4376, 4427, 4428, 4439, 4465,
    4551, 4582, 4618, 4620, 4621, 4660, 4669, 4670, 4721, 4722, 4760, 4832, 4895,
    4901, 4916, 4994, 5069, 5093, 5236, 5283, 5347, 5486, 5509, 5579, 5696, 5828,
    5882, 5908, 5969, 5992, 6009, 6019, 6025, 6027, 6032, 6043, 6136, 6223, 6249,
    6250, 6254, 6340, 6343

Water transport in plant, transport in xylem, methods 309, 729, 921, 1210, 1688,
    3013, 3022, 3048, 4155, 4524, 5988, 6314, 6532

Water transport in plant, transport in xylem of herbaceous stem 110, 184, 376, 515,
    841, 843, 913, 920, 921, 1075, 1096, 1181, 1299, 1492, 1562, 1613, 1620, 1736,
    1750, 1782, 2117, 2217, 2319, 2410, 2705, 2728, 2810, 3010, 3226, 3751, 3875,
    4091, 4155, 4287, 4572, 4573, 4662, 4728, 4832, 4940, 4976, 4985, 5069, 5240,
    5337, 5338, 5382, 5486, 5521, 5654, 5771, 5889, 5969, 5988, 6004, 6009, 6025,
    6032, 6297, 6314, 6340

Water transport in plant, transport in xylem of tree stem 294, 321, 376, 377, 651,
    729, 739, 843, 920, 1210, 1211, 1579, 1613, 1653, 1688, 1732, 1840, 1907, 2128,
    2130, 2205, 2410, 2412, 2470, 2540, 2630, 2637, 2639, 2665, 2810, 2852, 3022,
    3294, 3354, 3557, 3667, 3771, 3827, 3933, 4182, 4267, 4476, 4529, 4737, 5069,
    5237, 5239, 5302, 5382, 5519, 5520, 5879, 5969, 6009, 6025, 6181, 6254, 6297,
    6482, 6532

Water transport in plant, transport soil - root 62, 227, 312, 437, 449, 475, 569,
    605, 633, 659, 683, 714, 764, 851, 883, 1013, 1028, 1191, 1196, 1299, 1334, 1361,
    1450, 1542, 1563, 1620, 1695, 1850, 1859, 1863, 1865, 1898, 2003, 2201, 2258,
    2290, 2337, 2441, 2540, 2705, 2757, 2780, 2810, 2880, 2908, 3010, 3116, 3473,
    3475, 3557, 3967, 4124, 4183, 4287, 4465, 4660, 4895, 4916, 4994, 5069, 5389,
    5569, 5577, 5578, 5579, 5595, 5794, 5882, 5908, 5969, 6009, 6019, 6254, 6274,
    6343, 6440, 6537

Water transport in plant, vascular bundle structure 134, 170, 280, 292, 301, 316,
    367, 427, 441, 485, 522, 570, 671, 675, 676, 732, 743, 841, 842, 880, 913, 919,
    962, 974, 1043, 1073, 1152, 1208, 1472, 1515, 1517, 1616, 1656, 1739, 1861, 1971,
    2107, 2108, 2174, 2235, 2345, 2410, 2445, 2465, 2477, 2637, 2639, 2755, 2844,
    2858, 2880, 2893, 3010, 3046, 3102, 3352, 3400, 3543, 3751, 3816, 3955, 3962,
    4053, 4155, 4182, 4427, 4428, 4447, 4476, 4520, 4652, 4662, 4686, 4750, 4761,
    4872, 5084, 5093, 5118, 5124, 5149, 5237, 5239, 5240, 5247, 5535, 5771, 5951,
    6015, 6018, 6025, 6181, 6251, 6252, 6290, 6297

Wilting, effect on plant metabolism (continued) 2820, 2867, 2927, 2928, 2993, 3023,
    3050, 3096, 3130, 3212, 3275, 3291, 3312, 3358, 3408, 3409, 3517, 3661, 3671,
    3723, 3799, 3963, 4013, 4014, 4286, 4295, 4335, 4369, 4431, 4482, 4487, 4577,
    4689, 4745, 4748, 4792, 4802, 4804, 4849, 4899, 4946, 4975, 5002, 5023, 5024,
    5052, 5065, 5066, 5324, 5376, 5523, 5603, 5623, 5658, 5670, 5676, 5684, 5699,
    5760, 5887, 5908, 5912, 5931, 6309, 6360, 6393, 6492, 6497

Wilting, indicators, methods 1293, 1381, 2537, 2638, 2914, 3100, 3273, 3476, 3516,
    3592, 3606, 3605, 3640, 3722, 3987, 4061, 4090, 4211, 4697, 4784, 5161, 5678,
    6058, 6338, 6452

Wilting, mechanisms of development, indicators 63, 68, 69, 102, 133, 136, 156, 162,
    190, 276, 306, 308, 313, 317, 318, 391, 404, 470, 478, 498, 542, 561, 596, 666,
    679, 687, 688, 728, 767, 836, 899, 906, 920, 936, 955, 956, 960, 977, 1030, 1074,
    1088, 1124, 1130, 1143, 1264, 1266, 1268, 1297, 1304, 1343, 1393, 1395, 1400,
    1420, 1423, 1578, 1620, 1622, 1709, 1710, 1740, 1763, 1900, 1990, 2066, 2131,
    2148, 2149, 2200, 2201, 2214, 2243, 2293, 2320, 2328, 2335, 2362, 2390, 2397,
    2398, 2482, 2487, 2494, 2501, 2531, 2538, 2545, 2546, 2549, 2567, 2583, 2595,
    2652, 2655, 2679, 2684, 2695, 2710, 2787, 2801, 2804, 2806, 2807, 2831, 2892,
    2906, 2923, 2933, 2954, 2957, 2977, 2979, 2990, 3011, 3015, 3041, 3075, 3103,
    3104, 3107, 3130, 3131, 3132, 3142, 3153, 3167, 3231, 3276, 3278, 3292, 3293,
    3388, 3417, 3451, 3464, 3465, 3505, 3522, 3642, 3662, 3671, 3695, 3696, 3698,
    3701, 3710, 3712, 3764, 3788, 3789, 3799, 3800, 3809, 3819, 3820, 3866, 3868,
    3874, 3904, 3925, 3933, 3952, 3974, 3978, 3999, 4013, 4018, 4100, 4147, 4152,
    4213, 4218, 4219, 4268, 4270, 4271, 4306, 4307, 4335, 4347, 4348, 4356, 4363,
    4390, 4420, 4431, 4450, 4464, 4477, 4491, 4501, 4506, 4534, 4546, 4561, 4562,
    4574, 4577, 4594, 4607, 4610, 4636, 4644, 4661, 4664, 4667, 4670, 4672, 4678,
    4723, 4725, 4745, 4748, 4762, 4769, 4770, 4771, 4808, 4816, 4829, 4836, 4844,
    4849, 4874, 4875, 4895, 4911, 4927, 4956, 4964, 4965, 4997, 5002, 5018, 5025,
    5033, 5046, 5097, 5102, 5109, 5131, 5136, 5195, 5204, 5211, 5305, 5324, 5373,
    5376, 5378, 5384, 5385, 5388, 5406, 5413, 5430, 5486, 5495, 5532, 5533, 5551,
    5557, 5623, 5668, 5670, 5687, 5688, 5760, 5783, 5787, 5788, 5849, 5858, 5879,
    5891, 5896, 5910, 5912, 5916, 5946, 6006, 6024, 6025, 6032, 6034, 6044, 6122,
    6167, 6207, 6208, 6209, 6216, 6237, 6297, 6317, 6332, 6339, 6340, 6428, 6434,
    6438, 6440, 6469, 6514, 6515, 6526, 6568, 6583, 6587

Wilting, seasonal course 1143, 1263, 1343, 1356, 1508, 1789, 2598, 2933, 3296, 3551,
    3675, 3860, 4018, 4291, 4371, 4478, 6026, 6317

Wind, effect on conductance for water vapour and CO_2 transfer 279, 340, 608, 739,
    1102, 1180, 1409, 1444, 1445, 2381, 2632, 2795, 2842, 2843, 3405, 3977, 4110,
    4231, 4408, 4589, 4676, 4872, 4873, 5211, 5470, 5577, 5733, 5870, 6130, 6174,
    6246

Wind, effect on stomata and epidermis 808, 1045, 1783, 1802, 2632, 2843, 2844, 4317,
    4872, 5958, 5976

Wind, effect on transpiration 258, 279, 323, 545, 616, 758, 1045, 1056, 1102, 1147,
    1159, 1184, 1269, 1409, 1444, 1578, 1658, 1692, 1729, 1788, 2464, 2632, 2843,
    2886, 2933, 2951, 3005, 3620, 3665, 3862, 3977, 4110, 4548, 4997, 5056, 5211,
    5733, 5915, 6009, 6121, 6413, 6477, 6555

Wind, effect on water absorption by plant 2733, 3055, 3665, 6009

Wind, effect on water status in plant 526, 722, 808, 1142, 1802, 2200, 2276, 2795,
    2844, 2933, 3706, 3977, 4110, 4224, 4388, 4408, 4548, 4832, 4872, 4873

Wind, effect on wilting 1074, 1143, 2632

X

Xylem transport of water see Water transport in plant, transport in xylem of herba-
    ceous stem; Water transport in plant, transport in xylem of tree stem